APPLICATIONS AND DEVELOPMENTS OF BARODESY

ADVANCES IN GEOTECHNICAL ENGINEERING AND
TUNNELLING

24

General editors:

R. Hofmann, W. Fellin

University of Innsbruck, Division of Geotechnical and Tunnel Engineering

In the same series (A.A.BALKEMA):

1. D. Kolymbas (2000), *Introduction to hypoplasticity*, 104 pages, ISBN 90-5809-306-9

2. W. Fellin (2000), *Rütteldruckverdichtung als plastodynamisches Problem (Deep vibration compaction as a plastodynamic problem)*, 344 pages, ISBN 90-5809-315-8

3. D. Kolymbas & W. Fellin (2000), *Compaction of soils, granulates and powders - International workshop on compaction of soils, granulates, powders*, Innsbruck, 28-29 February 2000, 344 pages, ISBN 90-5809-318-2

In the same series (LOGOS):

4. C. Bliem (2001), *3D Finite Element Berechnungen im Tunnelbau (3D finite element calculations in tunnelling)*, 220 pages, ISBN 3-89722-750-9

5. D. Kolymbas, ed. (2001), *Tunnelling Mechanics, Eurosummerschool, Innsbruck, 2001*, 403 pages, ISBN 3-89722-873-4

6. M. Fiedler (2001), *Nichtlineare Berechnung von Plattenfundamenten (Nonlinear Analysis of Mat Foundations)*, 163 pages, ISBN 3-8325-0031-6

7. W. Fellin (2003), *Geotechnik - Lernen mit Beispielen*, 230 pages, ISBN 3-8325- 0147-9

8. D. Kolymbas, ed. (2003), *Rational Tunnelling, Summerschool, Innsbruck 2003*, 428 pages, ISBN 3-8325-0350-1

9. D. Kolymbas, ed. (2004), *Fractals in Geotechnical Engineering, Exploratory Workshop, Innsbruck, 2003*, 174 pages, ISBN 3-8325-0583-0

10. P. Tanseng (2006), *Implementation of Hypoplasticity for Fast Lagrangian Simulations*, 125 pages, ISBN 3-8325-1073-7.

11. A. Laudahn (2006), *An Approach to 1g Modelling in Geotechnical Engineering with Soiltron*, 197 pages, ISBN 3-8325-1072-9.

12. L. Prinz von Baden (2005), *Alpine Bauweisen und Gefahrenmanagement*, 212 pages, ISBN 3-8325-0935-6.

13. D. Kolymbas, A. Laudahn, eds. (2005), *Rational Tunnelling, 2nd Summerschool, Innsbruck 2005*, 291 pages, ISBN 3-8325-1012-5.

14. T. Weifner (2006), *Review and Extensions of Hypoplastic Equations*, 240 pages, ISBN 978-3-8325-1404-4.

15. M. Mähr (2006), *Ground movements induced by shield tunnelling in non-cohesive soils*, 168 pages, ISBN 978-3-8325-1361-0.

16. A. Kirsch (2009), *On the face stability of shallow tunnels in sand*, 178 pages, ISBN 978-3-8325-2149-3.

17. D. Renk (2011), *Zur Statik der Bodenbewehrung*, 165 pages, ISBN 978-3-8325-2947-5.

18. B. Schneider-Muntau (2013), *Zur Modellierung von Kriechhängen*, 225 pages, ISBN 978-3-8325-3474-5.

19. A. Blioumi (2014), *On Linear-Elastic, Cross-Anisotropic Rock*, 215 pages, ISBN 978-3-8325-3584-1.

20. G. Medicus (2015), *Barodesy and its Application for Clay*, 131 pages, ISBN 978-3-8325-4055-5.

21. C.-H. Chen (2015), *Development of Soft Particle Code (SPARC)*, 203 pages, ISBN 978-3-8325-4070-8.

22. I. Polymerou (2017), *Untersuchung großer Verformungen in der Vertushka*, 175 pages, ISBN 978-3-8325-4396-9.

23. D. Kolymbas (2018), *Spuren im Sand*, 97 pages, ISBN 978-3-8325-4683-0.

Applications and developments of Barodesy

Fabian Schranz
University of Innsbruck, Division of Geotechnical and Tunnel Engineering

E-mail: geotechnik@uibk.ac.at
Homepage: http://www.uibk.ac.at/geotechnik

Die ersten Bände dieser Reihe sind im Balkema Verlag erschienen. Bitte richten Sie Ihre Bestellungen von Band 1 bis 3 an folgende Adresse:

A.A. Balkema Publishers
P.O.Box 1675
NL-3000 BR Rotterdam
e-mail: orders@swets.nl
website: www.balkema.nl

Diese Publikation wurde mit finanzieller Unterstützung aus den Fördermitteln des Vizerektorats für Forschung der Leopold-Franzens-Universität Innsbruck gedruckt.

Bibliografische Information der Deutschen Nationalbibliothek

Die Deutsche Nationalbibliothek verzeichnet diese Publikation in der Deutschen Nationalbibliografie; detaillierte bibliografische Daten sind im Internet über http://dnb.d-nb.de abrufbar.

ISBN 978-3-8325-4883-4

ISSN 1566-6182

Logos Verlag Berlin GmbH
Comeniushof, Gubener Str. 47,
10243 Berlin
Tel.: +49 030 42 85 10 90
Fax: +49 030 42 85 10 92
INTERNET: http://www.logos-verlag.de

Acknowledgement

I am grateful to my supervisor Prof. Wolfgang Fellin for his invaluable support during the last years. Thank you for all the discussions, advice corrections and everything else.

I would also like to thank my second supervisor Prof. Dimitrios Kolymbas for suggestions, which improved the work.

I sincerely thank Prof. Erich Bauer for his useful feedback and his review activities.

Furthermore I to thank

- Dr. Gertraud Medicus for all the help and the many discussion about constitutive modelling and Dr. Barbara Schneider-Muntau for her support in writing scientific articles. It was always a great pleasure to collaborate with you two.

- Martin Schwarz and Matthias Rauter for their help with my math problems and their work in the students union.

- Sarah-Jane Loretz-Theodorine for her support and her sweets.

- my actual and former coworkers in the Unit for Geotechnical and Tunnel Engineering Iman Bathaeian, Manuel Bode, Christine Neuwirt, Claudia Thurnwalder, Stefan Tilg, Franz Berger, Dr. Anastasia Blioumi, Dr. Chien-Hsun Chen and Dr. Iliana Polymerou. It was great to work with you and I learned a lot of you.

- my flat mates Anna, Sebi and Edmound, which had no easy time with me in the last weeks.

Finally, I thank my family and friends for their personal support.

Kurzfassung

Barodesie ist ein Materialmodell für granulare Stoffe wie Sand und Ton. Das Modell basiert auf deren asymptotischen Verhalten bei konstanter Deformationsrate, das bedeutet, dass sich für eine konstante Deformationsrate die Spannung einem bestimmten Spannungsverhältnis annähert.

In dieser Arbeit wird die bestehende Sandversion der Barodesie weiterentwickelt. Dazu werden die Zugrundeliegenden skalaren Gleichungen verbessert und vereinfacht. Dies geschieht dadurch, dass verschiedene Konzepte aus der Bodenmechanik verwendet werden. Außerdem wird eine neue critical state line in das Modell integriert.

Die verbesserte Version wird im Anschluss mit unterschiedlichen elastoplastischen und hypoplastischen Materialmodellen verglichen. Dazu werden zum einen Laborergebnisse nachgerechnet und fortgeschrittene Spannungspfade simuliert (bei denen ausschließlich eine Rotation der Hauptnormalspannungen auftritt). Zum anderen werden Untersuchungen über die Standsicherheit unendlich langer Hänge durchgeführt, welche auf einen Simple Shear Versuch zurückgeführt werden können.

Abstract

Barodesy is a recently developed constitutive model for granular materials such as sand and clay. It is based on the asymptotic behaviour of granular media at a constant deformation rate. This means that for a constant deformation rate, the stress approaches a certain stress ratio.

In this work the existing sand version of Barodesy is improved. For this purpose, the underlying scalar equations are simplified and changed using different concepts from soil mechanics. In addition, a new critical state line is integrated into the model.

The improved version is then compared with different elastoplastic and hypoplastic constitutive relations. For this purpose, advanced stress paths (an exclusive rotation of the principal stresses) are simulated and compared with laboratory test results. In addition, investigations on the stability of an infinite slope is carried out, which can be traced back to a simple shear test.

Contents

List of Figures

List of Tables

Chapter 1

Introduction, overview and basic notation

Material models are the backbone of every modern finite element calculation. With the increase in computing capacity and the improvement of numerical methods, more and more finite element calculations are being carried out. In order to obtain realistic calculation results, advanced material models are required, especially in geotechnics.

The aim of research is to develop better and – maybe even more important – simpler material models. In this work, the material model Barodesy is improved. This recently developed material model differs fundamentally from elastoplastic constitutive models, since it requires neither yield surfaces, plastic potentials nor hardening laws. Barodesy can be written in one equation, what makes it similar to Hypoplasticity.

1.1 Overview

This thesis summarises my research on and with constitutive models, which I have undertaken in the last four years at the University of Innsbruck during my PhD studies.

The second chapter consists of the paper "Zur Rolle der Materialmodelle beim Standsicherheitsnachweis" from Kolymbas *et al.* [65] in geotechnik. It is mainly concerned with the fundamental question of whether the material model is important for the determination of stability or not. Statements of the Working Group for Numerics in Geotechnics are discussed in detail and compared with calculation results from the literature.

The third chapter takes a closer look to the basic behaviour of soil and presents the material model Barodesy. The ability of Barodesy to depict soil behaviour is explained in more detail. At the end of this chapter, the two current versions of Barodesy for clay and sand are presented. This chapter is based on the publication "Konzepte der Barodesie" published by Medicus *et al.* [78] in Bautechnik.

In the fourth chapter, improvements of Barodesy for sand are developed. On the one hand, the existing formulation of a kernel function will be improved and simplified, and on the other hand, another critical state line will be introduced, which requires to change some scalar functions of Barodesy.

In the fifth chapter, advanced stress paths (principal stress rotation) with different material models (different elastoplastic, hypoplastic and barodetic material models) are calculated and the results of the calculations compared with results of laboratory tests. These results were also included in the paper "Deformations induced by principal stress rotation modelled with different constitutive relations" (submitted to the International Journal for Numerical and Analytical Methods in Geomechanics without the improved Barodesy version and without a detailed presentation of the material models).

The sixth chapter deals with the stability of infinite slopes. Different approaches for calculating the stability of infinite slopes are presented. Various material models are used for the calculation of failure in simple shear simulations. It is possible to derive simple formulas for certain special cases. A large part of this work is published in "Stability of infinite slopes investigated with Elastoplasticity and Hypoplasticity" by Schranz and Fellin [106] in geotechnik. The here presented work is extended by calculations with Barodesy, which were not included in the original.

Beside of the work on constitutive models there has been also further research on the implementation of constitutive models into Finite Element software Chen *et al.* [15] and the estimation of material parameters under the aspect of the safety concept in the new standardisation Schneider-Muntau *et al.* [104].

1.2   Basic notation

1.2.1   Stress, strain and stretching

In this thesis the effective Cauchy stress Tensor is denoted as $\boldsymbol{T}$, following the sign convention of the continuum mechanics (tension positive). In the case of principal normal stresses, the components of the stress tensor are written with a single index, for the smallest stress component (the largest absolute value) carries the index 1, the largest stress component (the smallest absolute value) the index 3 ($T_1 \leq T_2 \leq T_3$). If a general stress state is described, two indices are used: T_{ij}. In the case of $i = j$, these are normal stresses and for $i \neq j$ these are shear stresses.

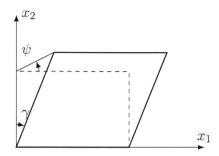

Figure 1.1: Deformation in a simple shear test

The strain $\boldsymbol{\varepsilon}$ also follows the sign convention of continuum mechanics (elongation positive). All constitutive models used in this thesis are of the rate type. Hence, in the following the rate of deformation tensor (also called stretching) $\boldsymbol{D}$ is used, which is the symmetrical part of the velocity gradient. In the case of rectilinear deformations and the use of the logarithmic strain, $\boldsymbol{D} = \dot{\boldsymbol{\varepsilon}}$ applies (Gurtin and Spear [49]).

1.2.2   Dilatancy

Dilatancy is something very characteristic for granular media like soil and was first scientifically described by Reynolds [95], although the effect was already known earlier.

In the case of granular media, dilatancy is the change of volume during a shear deformation. Different dilatancy measures are common in soil mechanics, they are partly closely related to the constitutive model which is used. Due to the importance of dilatancy and its frequent occurrence in this work, as well as its different definitions, it should be discussed in more detail in this section.

The simplest case, where the dilatancy can be seen, is in a plain strain simple shear test (cf. Fig. 1.1). In this test the horizontal normal strains are constraint and only vertical normal strain and shear strain can occur. The dilatancy angle ψ is then defined as

$$\tan \psi = \frac{\mathrm{d}x_2}{\mathrm{d}x_1} = \frac{D_{22}}{2D_{12}} \quad . \tag{1.1}$$

In elastoplastic formulations dilatancy is related to plastic flow. Therefore the plastic stretching $\boldsymbol{D}^{\mathrm{p}}$ is used with

$$\boldsymbol{D} = \boldsymbol{D}^{\mathrm{e}} + \boldsymbol{D}^{\mathrm{p}} \tag{1.2}$$

following the decomposition of strain in an elastic and plastic part. That yields

$$\tan \psi = \frac{D^{\mathrm{p}}_{22}}{2D^{\mathrm{p}}_{12}} \quad . \tag{1.3}$$

For general deformations, the equation for the dilatancy angle for an elasto-plastic model is defined as

$$\sin \psi = \frac{\operatorname{tr} \boldsymbol{D}^{\mathrm{p}}}{|D^{\mathrm{p}}_1 - D^{\mathrm{p}}_3|} \quad . \tag{1.4}$$

Here the maximum and the minimum principal stretching are used. For the triaxial test this equation results in

$$\sin \psi = \frac{\operatorname{tr} \boldsymbol{D}^{\mathrm{p}}}{-2D^{\mathrm{p}}_1 + \operatorname{tr} \boldsymbol{D}^{\mathrm{p}}} \quad , \tag{1.5}$$

with the axial plastic stretching D^{p}_1. The definition of the dilatancy as a function of plastic deformations is problematic, since plastic strains cannot be measured during a laboratory test (before unloading) and it is not applicable for constitutive relations without plastic strain (e.g. Hypoplasticity or Barodesy). This problem is sometimes overcome with the assumption that the entire deformation is plastic and hence $\boldsymbol{D} = \boldsymbol{D}^{\mathrm{p}}$. This assumption holds true only in the critical state.

Chu and Lo [16] use a different measurement for the dilatancy. They use $\tan \beta$ which is just defined for axisymmetric states and reads

$$\tan \beta = \frac{-\operatorname{tr} \boldsymbol{D}}{D_1} \quad . \tag{1.6}$$

A further possible measure for dilatancy, which is more general, is

$$\delta = \frac{\operatorname{tr} \boldsymbol{D}}{\|\boldsymbol{D}\|} \quad , \tag{1.7}$$

where the trace of the stretching $\boldsymbol{D}$ and its absolute value $\|\boldsymbol{D}\| = \sqrt{\operatorname{tr} \boldsymbol{D}^2}$ are used. The range of δ is between $-\sqrt{3}$ for hydrostatic compression and $\sqrt{3}$ for hydrostatic extension. The relation between $\tan \beta$ and δ can be calculated for axisymmetric states as

$$\delta = \frac{\tan \beta}{\sqrt{1 + \frac{(1+\tan \beta)^2}{2}}} \quad . \tag{1.8}$$

This relation is shown in Fig. 1.2

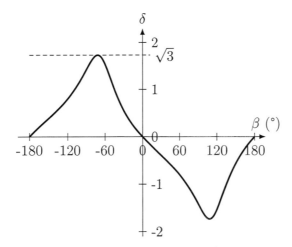

Figure 1.2: Relation between the two dilatancy measurements $\tan \beta$ and δ

1.2.3 Rendulic plane

The Rendulic plane is a plane in the principal stress/strain space, which was introduced by Rendulic [94]. This plane includes the hydrostatic stress/strain axis ($T_1 = T_2 = T_3$ or $D_1 = D_2 = D_3$), Fig. 1.3a. This plane has the advantage that axisymmetric stress or strain paths (for which is $T_2 = T_3$ or $\varepsilon_2 = \varepsilon_3$) are shown undistorted, for this reason the value on the horizontal axis (T_2 or D_2) is scaled by the factor $\sqrt{2}$ (cf. Fig. 1.3). Following Gudehus and Mašín [47], in this plane the angles ψ_T and ψ_D can be defined. For isotropic compression ψ_T and ψ_D are defined to be zero, so the angles are

$$\psi_T = \arctan \frac{T_1}{\sqrt{2}T_2} - \arctan \frac{1}{\sqrt{2}} \quad , \tag{1.9}$$

$$\psi_D = \arctan \frac{D_1}{\sqrt{2}D_2} - \arctan \frac{1}{\sqrt{2}} \quad . \tag{1.10}$$

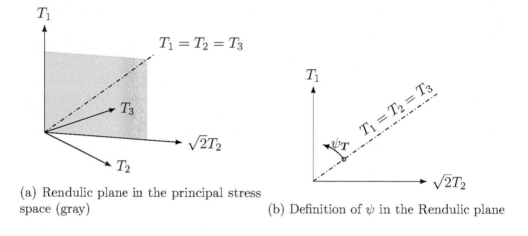

(a) Rendulic plane in the principal stress
space (gray)

(b) Definition of ψ in the Rendulic plane

Figure 1.3: Rendulic plane in the stress space (it is also valid for strains)

Chapter 2

The role of constitutive models

The stability of a geotechnical building is often assumed to be not very sensitive to the choice of the constitutive model according to the Empfehlungen des Arbeitskreises für Numerik in der Geotechnik – EANG [35] from the German Geotechnical Society section 2.6.3.2 "Stoffmodelle für Standsicherheitsberechnung". However, this statement can mislead to wrong conclusions or incorrect use of soil models, if an engineer is not experienced in this field. This chapter investigate possible limits of this statement. Commonly it is emphasised, that nowadays the constitutive models plays an important role, since the computing capacity and the numeric methods are sophisticated. The EANG also mentions, that the typical elastoplastic model with a Mohr-Coulomb yield-surface (in the following MC-Model) is sufficient. Together with the information, that for excavations and slop stability analysis the dilatancy is negligible, because of the small restraint of the deformation[1] [35, p. 60] (see also Davis [21]), this can mislead someone to use the MC-Model unconsidered.

A major reason for the great acceptance of such oversimplified statements may be, that there is a long tradition of a dichotomy between stability analysis and the deformation problems in geotechnics. In the beginning just stability analysis was considered in geotechnical engineering (using a somehow arbitrary defined level of safety), relatively late also deformation problems were encountered. Always with the knowledge, that the accuracy of the deformation prediction is quite lower than the one of the stability calculation. This dichotomy is also still consistently implemented in teaching.

At the beginning the research focused on the event of the "failure" or "fracture" and tried to describe it with various strength hypotheses. For a long time deformations have been calculated using elasticity theory. With the introduction of the Elastoplasticity theory one began to realise, that failure is nothing different than a special form of a deformation state. Furthermore, Elastoplasticity theories insinuate that failure occurs precisely when a certain stress state is reached. It is ignored that the failure of solids is rather a process

---

[1] The original Phrase in [35] reads: Für Baugruben und Böschungen zeigt sich auf Grund der relativ geringen Behinderung der Verformungen nur ein vernachlässigbarer Einfluss des Dilatanzwinkels auf die Standsicherheit,...

than a state, which is difficult to determine, as it is also associated with the breakdown of numerical methods and the loss of controllability of experimental methods. An expression of "trivialisation" of failure is the term of the equilibrium equation, which found its way into the standardisation. The trivialisation consists in the belief that failure can be described by an "algebraic equation". This may be possible for states, but not for processes. By the way, the fact that failure is more a process than a singular event was already taken into account by Terzaghi and Peck [113] with the introduction of the concept of the progressive breakage, which was hardly followed up.

So called *strength reduction methods* are used in finite element calculations. The values of the used friction angle φ and the cohesion c are reduced until a limit state is reached in the case of φ-c-reduction. This results in localisation of deformation in a thin area, which can be compared with "traditional" sliding circles. Compared to kinematic methods (like sliding circles or rigid-body failure mechanisms), the finite element methods offers a quite good fulfilment of the global equations of forces and momentum. However, it must be emphasised that the application of strength reduction methods do not always deliver unambiguous results. The safety obtained with such methods depends strongly on the concrete application of the reduction and on the individual material parameters, (c.f. Fellin [29] and Zhao *et al.* [132]). In any case, the φ-c-reduction shows that the aforementioned dichotomy between deformation calculation and stability analysis fails. With a constitutive model that is unsuitable for the problem, an incorrect stress path to the limit state is obtained and therefore also inaccurate stress states at the failure are obtained. This applies in particular to undrained conditions and was one of the reasons for the overestimation of the shear strength at the accident next to the Nicoll Highway (see section 2.3.2). Moreover, the φ-c-reduction can only be used in models, which explicitly contain these parameters, such as the Mohr-Coulomb model. For the case that other constitutive models are used different methods have to be applied. For example, Schneider-Muntau *et al.* [103] reduce the critical state friction angle φ_c and the specific volume parameter of the critical state line N.

2.1   About the Failure

In the following, failure should be limited to failure of soil samples in laboratory tests, especially to triaxial tests. From a phenomenological point of view, failure manifest itself in a horizontal tangent of the stress-strain curve (a so-called limit state). What is quite often missed is that neither stresses nor strains can be measured directly, it is only possible to measure displacements and forces. In order to deduce stresses and strains from these measurements, one needs a substantial assumption that the deformation of the specimen is homogeneous, i.e. that it retains its cylindrical shape. However, experience has shown that near the limit state the deformation get inhomogeneous, the specimen is deformed unevenly, it bugles out or it is pervaded by shear bands. For a while it was believed that the loss of homogeneity was caused by disturbing boundary effects. It was therefore attempted to eliminate the friction at the end of the sample by means of lubrication and/or to make the sample more compact, but it was found that inhomogenization could not be avoided. Later, the theory has shown that internal inhomogenization is unavoidable (see section 2.1.2), because at some point of the experiment the so-called controllability gets lost, i.e. it is not possible to force the distribution of stresses and deformations *within* the sample by applying stresses and displacements to the boundary of the sample. If the controllability is lost, the sample – loosely formulated – can decide for itself which deformation it will undergo. The specimen often "choose" a localised deformation that takes place within thin shear bands. The appearance of shear bands is a kind of pattern formation in an originally uniform sample. The possibility of inhomogeneous sample deformation as an alternative to the homogeneous deformation which has been introduced so far is accompanied by the loss of uniqueness of the solution of the underlying problem of the initial boundary value problem and by a bifurcation of the solution path, whereby the solution found thereafter, e.g. in a finite element calculation, is mesh dependant. So it can be seen that the closer one get to the limit state, the less informative the experiments become. The bifurcation of the deformation manifests itself in the stress-strain-curve in the sense, that it cannot be traced to the peak by a laboratory test, as it is already distorted by the occurrence of the inhomogeneous deformation.

Failure can be seen as a kind of phase transition that begins at a single nucleus (such as a small imperfection) and transform the material from a solid to a material that is able to flow (but only in one direction). This concept has already been adopted in fracture mechanics and also puts a basic assumption of our standard numerical simulations in question. This is the assumption of the simple material, which states that the size of a specimen does not play a role in its stress-strain-behaviour. In fact it is observed that larger samples have a lower strength.

Deformations, which can be small or large, are closely linked to failure. Whether these deformations occur abruptly or slowly, i.e. if the material behaves ductile or brittle, is another question.

2.1.1   The role of dilatancy at failure

The EANG [35] suggests that dilatancy only plays a role in the case of re-straint deformations. However, it should be noted that the difference between the peak friction angle and the critical friction angle is due to dilatancy, what was pointed out by Taylor as first [112]. This means that dilatancy is always important when the peak friction angle is of importance and not the criti-cal fiction angle. Since the dilatancy is suppressed with increasing pressure level, the peak friction angle is strongly stress-dependant (this effect is known as barotropy). Hence, the peak friction angle can not be seen as a material constant, as suggested by the Mohr-Coulomb failure criterion. This implies that the Mohr-Coulomb model is not the best choice for problems with highly variable stress level (e.g. piles, ground foundation, penetration [18]) .

2.1.2   Mathematical-mechanical description of failure

As we have seen above, it can happen that the limit state cannot be reached in an experiment, because a bifurcation occurs before the limit state is reached and the sample deforms inhomogeneous. This means that failure does not occur by reaching the limit state, but due to inhomogeneous deformations. The underlying mathematical event is the loss of the uniqueness of the solution to the considered initial boundary-value problem. This can also be seen as a loss of controllability of the sample deformation by specifying displacements and/or stresses at the boundary of the specimen. Nova [89] has shown that this is associated with the vanishing of the so-called second order work. This is the value

$$\mathrm{d}^2 W = \mathrm{d}T_{ij}\,\mathrm{d}\varepsilon_{ij} \ \text{ or} \tag{2.1}$$

$$\mathrm{d}^2 W = \dot{T}_{ij} D_{ij} \tag{2.2}$$

which can be regarded as the second variation of the deformation work

$$\int T_{ij} D_{ij}\,\mathrm{d}V \tag{2.3}$$

on a volume element $\mathrm{d}V$. Whether $\mathrm{d}^2 W$ can vanish while performing an el-ement test can be investigated with a constitutive relation for a given stress state. For this purpose, it is necessary to calculate the associated stress incre-ments $\mathrm{d}T_{ij}$ for all possible strain increments $\mathrm{d}\varepsilon_{ij}$ with the constitutive relation and determine whether their product is positive. If this is not the case, an inhomogeneous deformation can occur.

2.1.3   Initial boundary value problems

Homogeneous deformation fields are not expected for general initial boundary value problems. The loss of uniqueness in the calculation leads to problems that have not yet been solved satisfactorily. In numerical calculations, e.g. with finite element methods, such a loss of uniqueness is manifested, among other things, by the fact that the stiffness matrix can be singular, i.e. no unique solution of the linear system of equations of the problem can be determined. The introduction of additional conditions may restore uniqueness. The methods used for this purpose (such as Cosserat continua) are called *regularisation* and represent further modelling assumptions, some of them are difficult to justify.

2.2   Validation of numerical simulation of failure in geotechnics

2.2.1   Measurements

Geotechnics can be compared with meteorology. In both disciplines mathematics plays an important role, but they still yield rather inaccurate results. It is not possible to predict the settlement of building with millimetre accuracy, and the weather forecast does not always coincide with the actual weather. And yet there is a big difference: While in meteorology predictions are validated (i.e. verified) on a daily basis, the calculations of geotechnical engineers are verified by other geotechnical experts, hardly ever they are verified by a comparison with reality. This applies in particular to limit loads, i.e. stability analysis. The aim of usual calculations in geotechnics is to proof that there is a still a sufficient safety against failure, and everyone is glad if the construction does not fail. Of course, there are cases in geotechnical engineering where a failure of the soil is aimed at, e.g. in the case of lowering a caisson or in penetration. Usually constructions are not designed to test if they fail and it is obvious that this approach is not appropriate for construction practice. Of course, mathematical-mechanical-numerical methods are used to determine the current soil parameters. However, this cannot be regarded as validating these methods.

2.2.2   Comparison between different calculations

What will then remain is to compare the results of numerical simulations with analytical solutions or traditional solutions (e.g. using a sliding circle method).

Analytical solutions include also the earth pressure according to Coulomb [17] and the bearing load of a foundation according to Prandlt [93]. Such solutions are widely accepted in geotechnical engineering. However, they are extremely difficult to validate but are still seen as equivalent to reality. If we take a look at the bearing capacity of a foundation according to Prandtl, we see that it increases exponentially with $\tan \varphi$. This means that a slight change in the friction angle φ has a very high impact on the bearing capacity. However, the friction angle cannot be measured with absolute accuracy, simply because it depends strongly on the pressure level. Which pressure level has to be used for the calculation of the bearing capacity? The large scale tests carried out by the Degebo on bearing capacity of foundations remain rather inconclusive with regard to the validation of the various analytical formulas proposed for this purpose [87]. The indecisiveness and scattering of measurements is also discussed in the work of Canepa and Garnier [14]. Also the earth pressure formulas can hardly be validated by experiments, if one considers that the angle of the wall friction plays a major role, but is usually applied by experience (about $2/3\varphi$) and has hardly ever been measured. Small-scale test at 1 g or in a centrifuge also have interpretation problems (e.g. the role of the grain diameter in shear joints or the measurement of the friction angle at very low stress levels).

Methods based on the sliding circle method or rigid body movements are very common in soil mechanics. These are applications of the kinematic load-bearing theorem of plasticity theory. Another approach uses limit stress fields according to the method of the characteristics, which are an application of the static load-bearing theorem. Whether the estimation of the limit load is on the safe side determined in this way or not, is another question, which depends on the validity of the so-called normality condition of plasticity theory.

The result of an analytical calculation is often compared with a numeric simulation. This may contravene the scientific theory prohibition to validate a theory on the basis of another theory. Cf. the second Feuerbach thesis by Marx:

> 'Die Frage, ob dem menschlichen Denken gegenständliche Wahrheit zukomme, ist keine Frage der Theorie, sondern eine praktische Frage. In der Praxis muss der Mensch die Wahrheit, d. h. die Wirklichkeit und Macht, die Diesseitigkeit seines Denkens beweisen. Der Streit über die Wirklichkeit oder Nichtwirklichkeit eines Denkens, das sich von der Praxis isoliert, ist eine rein scholastische Frage.' [26][2]

---

[2] The question if objective truth is possible to human thought is not a theoretical but

2.3. Influence of constitutive model

However, such comparisons often deliver satisfactory results (if they are consistent) which is seen as confirmation of the numerical approach. What is the reason for the coincidence? The reason is relatively simple. The shear strength parameters (friction angle φ and cohesion c) are used as input parameters in both procedures. If this is done, the validity of the failure hypothesis according to Mohr-Coulomb is presupposed – even if only tacitly – and the theory is exaggeratedly formulated by comparison with itself. It should therefore not be surprising that both methods produce approximately the same results.

For example, the EANG [35] compares the results of finite element calculations of an excavation with different constitutive models in the appendix (starting from page 115). A fictitious excavation is simulated for four different soils (dense sand, loose sand, soft soil and over-consolidated clay), each one with three constitutive models (an elastoplastic model with a yield condition according to Mohr-Coulomb, Hardening Soil and Hardening Soil Small). For the clayey soils also the Soft soil model is used. The differences in the deformation behaviour are significant, but the result of the stability analysis obtained by the shear strength reduction is comparable according to EANG [35]. In this context, it should be mentioned that all four models use the Mohr-Coulomb failure criterion as a yield surface. This explains the small differences in the stability analysis.

Also the so-called Finite Element Limit Analysis [109] is basically restricted to the Mohr-Coulomb strength hypothesis, since the Mohr-Coulomb parameters are reduced.

2.3 Influence of constitutive relations on the limit state

In this section, some examples are given in which the choice of the constitutive model influences the load bearing capacity.

2.3.1 Infinite slope

The statement of the EANG, that for excavations and slopes, only a negligible influence of the dilatancy angle on stability is evident due to the relatively low level of deformation [35], cannot be comprehended even for the simplest of all

---

a practical question. In practice man must prove the truth, that is the reality and force in his actual thoughts. The dispute as to the reality or non-reality of thought which separates itself, "the praxis," is a purely scholastic question. Translation from [27]

stability analysis, the infinite slope. If the limit state of an infinite slope is calculated for an elastoplastic material and a yield criterion according to Mohr-Coulomb, it follows from the formulation of the yield surface and the plastic potential that the maximum slope angle ς is a function of the friction angle φ and the dilatancy angle ψ (c.f. Teunissen and Spierenburg [114])

$$\tan\varsigma = \frac{\sin\varphi\cos\psi}{1-\sin\varphi\sin\psi} \quad . \tag{2.4}$$

Just in the case of an associated plastic potential (i.e $\psi = \varphi$) (2.4) reduces to the commonly used relation

$$\varsigma = \varphi \quad . \tag{2.5}$$

In the case of $\psi = 0$ (what complies to an loose sand) (2.4) reduces to

$$\tan\varsigma = \sin\varphi \quad , \tag{2.6}$$

with a stress state which is experimentally proven by Budhu [12] and numerically well documented by Thornton and Zhang [115, 116]. Further investigations of the stability of infinite slopes, also with other constitutive relations, are made in chapter 6.

2.3.2   Undrained conditions

Undrained condition represent a limitation of the deformations, and are thus excluded from the relevant statement of the EANG, that for excavations and slopes, only a negligible influence of the dilatancy angle on stability is evident due to the relatively low level of deformation. But this can easily be overlooked. The incorrect use of the elastoplastic constitutive model with a yield surface according to Mohr-Coulomb leads to a considerable overestimation of strength, especially in loose or normal-consolidated soils. The following examples illustrate the problem.

Nicoll Highway accident in Singapore

A good example of an overestimation of strength is the Nicoll Highway accident in Singapore in April 2004. During the construction of an underground express train line in a fine-grained soil the retaining wall failed at an excavation depth of about $30\,\mathrm{m}$ below ground level. The numerical calculations were carried out with effective shear parameters (φ and c) determined by a consolidated undrained triaxial test. The slightly permeable soil was modelled in a so-called undrained calculation, see Figure 2.1b, assuming that the water

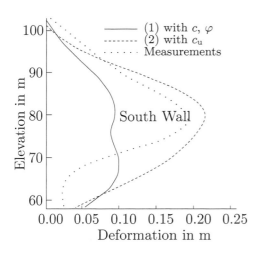

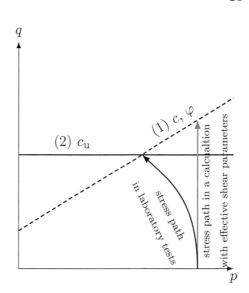

(a) Calculated and measured wall deformation for the Nicoll Highway accident at the last excavation step, shown are the results from Whittle and Davies [127] for a calculation with effective shear parameter c and φ and results with depth-dependent undrained shear strength c_u. The measured deformations according to Hight *et al.* [53] are also plotted for comparison.

(b) The shear strength for undrained conditions with the effective shear parameters c and φ (1) and a yield surface according to Mohr-Coulomb, compared to shear strength determined with the undrained shear strength c_u (2), compare Fig. B.1b. Starting from the same stress level, the shear strength for normally consolidated soil is overestimated.

Figure 2.1: Influence of the selected model for calculating undrained conditions on the wall deflection of an excavation pit wall and shear strength.

in the pores cannot flow out due to the low permeability of the soil (which imposes a constant volume deformation). In a calculation with the elasto-plastic constitutive model with a yield surface according to Mohr-Coulomb, this means that the mean stress p remains constant until the yield surface is reached and only the deviatoric stress q increases, compare Appendix B. The wall deformations were severely underestimated with the chosen calculation method, cf. Fig. 2.1a. According to Whittle and Davies [127], various reasons for the damage were identified, including the choice of an unsuitable geotechnical calculation method. Recalculations of the excavation with an undrained, depth-dependent shear strength c_u resulted in considerably larger and more realistic wall deformations. Figure 2.1a shows the wall deformation predicted with the two calculation methods, as well as the measured wall deformations according to Hight *et al.* [53].

Considerations in the deviatoric plane

The choice of the yield surface may have a major influence on stability calculations for general conditions. Figure 2.2a shows different yield surfaces in the deviatoric plane (i.e. $p = $ const). Experimental results of San Francisco Bay Mud from Lade [67] are shown. For axially symmetric triaxial compression, the predictions of all yield surfaces agree (i.e. $f_y = 0$). However, for general conditions there are significant differences.

Study case embankment

Grammatikopoulou *et al.* [42] investigate the influence of the yield function f_y and the plastic potential g_y on the failure height of an embankment. The embankment is modelled with an elastoplastic model with yield surface according to Mohr-Coulomb. The underground consists of clay deposits, which are modelled with the *two-surface "bubble" model* from Al-Tabbaa and Wood [1]. Grammatikopoulou [41] extended the model with the possibility to use different combinations of yield functions and plastic potentials for simulations with finite element program ICFEP. For their calculations, Grammatikopoulou *et al.* [42] choose combinations of f_y and g_y according to Table 2.1. The different combinations of the yield function and the plastic potential lead to very different calculation predictions, see Table 2.1. Case 3 provides the highest failure height, it is 41 % higher than in case 1, the failure height for case 2 is about 30 % higher than in case 1. The maximum mobilised friction angle is the highest in case 2, but the failure level is the largest in case 3. It is interesting to note that the failure height in case 2 and case 3 differ about 8 %, although the calculation only differs in the choice of plastic potential.

2.4   When do material models mater?

The statement in the EANG [35] on the role of constitutive relations in determining stability is relevant in many cases, however, there are some limitations, which have been shown here. The current state of knowledge still does not allow to give a final answer to the question of the correct constitutive model, its calibration, its numerical implementation, its eventual regularisation and interpretation as done. Therefore, it is always necessary to specify the constitutive model and the material parameters used. A sophisticated numerical calculation is carried out when important details such as the influence of porosity and pressure level are involved. These effects are not taken in account by a conventional elastoplastic model with a yield surface according

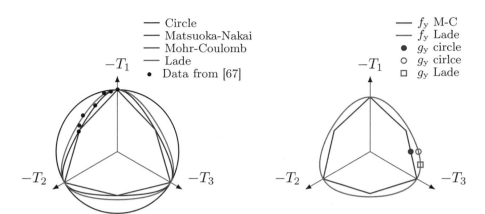

(a) Intersections of different yield surface with the deviatoric plane ($p \approx 167\,\text{kPa}$). Failure points of San Francisco Bay Mud ($\varphi_c = 30.6°$) from Lade [67] are also shown.

(b) Calculation of embankment results of the failure points on the corresponding yield surface. Figure slightly modified after Grammatikopoulou et al. [42].

Figure 2.2: Different yield surfaces in the deviatoric plane.

Table 2.1: Different combinations of yield function f_y and plastic potential g_y and their effect on the calculated results, from [41], see also Fig. 2.2b.

Case	yield function f_y	plastic potential g_y	failure height	max φ_{mob}
1 ●	Mohr-Coulomb	circle	3.95 m	27.0°
2 ∪	Lade	circle	5.12 m	32.7°
3 □	Lade	Lade	5.55 m	31.3°

to Mohr-Coulomb. In these cases advanced constitutive models have to be used, for example Barodesy (which is described in chapter 3), Hypoplasticity or Sanisand (which are used in chapter 5). If the influence of the stress level and the void ratio can be neglected, an analysis of fracture mechanisms (such as sliding circles or rigid body movements) may be sufficient.

Chapter 3

Barodesy

Barodesy is a constitutive model that differs fundamentally from conventional elastoplastic models. It does not distinguish between elastic and plastic strains and thus does not require notions such as a yield surface, plastic potential or flow rule. The objective stress-rate $\overset{\circ}{T}$ is a tensorial function of the actual stress T, the deformation rate D, which is in the case of rectilinear deformations the same as the logarithmic strain rate $\dot{\varepsilon}$, and the void ratio e. Barodesy has certain similarities with Hypoplasticity. In contrast to the Hypoplasticity, which is based on a representation theorem for tensorial functions[1], the basic structure of Barodesy is derived from the so-called *rules of Goldscheider* (cf. section 3.2.2).

Theories from soil mechanics (such as critical state soil mechanics) can be easily integrated in the framework of Barodesy. The following soil properties will be discussed under a barodetic view in the following:

1. soil is non-linear, the stiffness of soil is stress dependant

2. soil has an asymptotic behaviour for constant strain rates, see Goldscheiders rules (sec. 3.2.2)

3. soil has critical states, where the stiffness and volume strain rate vanish under ongoing shearing

4. soil has a stress-dilatancy relation: the higher the dilatancy the higher the peak friction angle

5. soil behaviour exhibits a dependence on density, called pyknotropy.

In section 3.2, the soil behaviour will be described in detail and results from laboratory tests will be shown.

---

[1] The Cayley-Hamilton theorem for tensorial functions of a tensorial argument states: A tensorial function $F(X)$ can be represented as the sum of the powers of X^0, X^1, X^2, and functions a_i of the invariants of X, i.e. $F(X) = a_0 I + a_1 X + a_2 X^2$, see Fellin and Kolymbas [32]

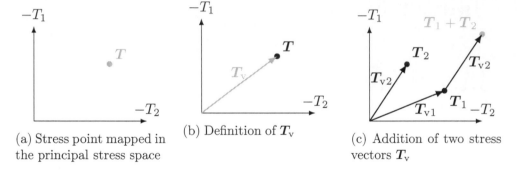

(a) Stress point mapped in the principal stress space

(b) Definition of $\boldsymbol{T}_\text{v}$

(c) Addition of two stress vectors $\boldsymbol{T}_\text{v}$

Figure 3.1: Relation between stress tensor $\boldsymbol{T}$ and stress vector $\boldsymbol{T}_\text{v}$ in the principal stress space

3.1 Notations used in Barodesy

Stress representation in principal stress space The stress tensor $\boldsymbol{T}$ can be mapped into the principal stress space as a point, Fig. 3.1a. A vector $\boldsymbol{T}_\text{v}$ can be defined, which points from the origin of the principal stress space to the stress point $\boldsymbol{T}$, Fig. 3.1b. The addition of two stress vectors $\boldsymbol{T}_{\text{v}1} + \boldsymbol{T}_{\text{v}2}$ points to the stress point resulting from $\boldsymbol{T}_1 + \boldsymbol{T}_2$, Fig. 3.1c.

Norm of a tensor $\|\boldsymbol{X}\|$ is defined as the Euclidean Norm $\|\boldsymbol{X}\| = \sqrt{\operatorname{tr} \boldsymbol{X}^2}$ (also called Frobenius norm). Note, the value of the norm of the stress tensor $\|\boldsymbol{T}\|$ is equal to the value of the norm of the stress vector $\|\boldsymbol{T}_\text{v}\| = \sqrt{T_1^2 + T_2^2 + T_3^2} = \|\boldsymbol{T}\|$.

Normed tensor $\boldsymbol{X}^0$ is defined as the tensor divided by his norm $\boldsymbol{X}^0 = \boldsymbol{X}/\|\boldsymbol{X}\|$. The normed stress tensor $\boldsymbol{T}^0$ can be interpreted as a tensor with the direction of the stress, because its length is 1 and it points into the direction of the stress tensor (see Fig. 3.2 for a graphical interpretation).

Mobilised friction angle From the failure criterion according to Mohr-Coulomb a mobilised friction angle φ_mob can be derived as

$$\sin \varphi_\text{mob} = \frac{T_3 - T_1}{T_1 + T_3} \quad . \tag{3.1}$$

After a long shearing the mobilised friction angle φ_mob is the same as the critical friction angle φ_c, for dense samples or dilatant strain paths the maximum of the mobilised friction angle can also be higher than the critical friction angle. The theoretical maximal mobilised friction angle φ_max is 90°.

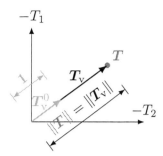

Figure 3.2: Normed stress vector $\boldsymbol{T}_\mathrm{v}^0$ in the principal stress space

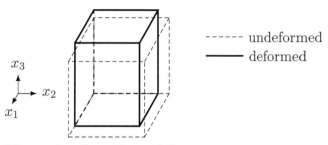

Figure 3.3: Rectilinear deformation

Proportional paths are stress or strain paths for which $T_i : T_j = $ const or $D_i : D_j = $ const is valid respectively. For example, the isotropic compression results in a proportional strain path with $D_1 = D_2 = D_3$ and so $D_1 : D_2 = 1$, $D_1 : D_3 = 1$ and $D_2 : D_3 = 1$. The corresponding stress path is also a proportional path, it is the path along the hydrostatic axis the ratios between the stresses are $T_1 : T_2 = 1$, $T_1 : T_3 = 1$ and $T_2 : T_3 = 1$. An other proportional path can be achieved in the oedometer test ($D_2 = D_3 = 0$), where the ratio between $T_2 : T_1 = T_3 : T_1$ results in the K_0-path. In the following, the direction of a proportional stress path is denoted as $\boldsymbol{R}^0$.

Rectilinear deformations are deformations at which a cuboid remains rectangular (cf. Fig. 3.3). For such deformations, the unsorted directions of the principal stress and principal strain remain the same.

3.2   Soil behaviour and Barodesy

In this section, soil properties and theories from soil mechanics are explained. Furthermore, it will be shown how these concepts can be easily taken into account by Barodesy.

3.2.1   Non-linear behaviour

The soil is characterised by a non-linear material behaviour. The stiffness of the soil depends in the stress level and the density (void ratio e). The higher the stress level and the density (smaller void ratio), the stiffer the soil will behave with given deformation $\boldsymbol{D}$. In the following, a hydrostatic compression test ($-p = T_1 = T_2 = T_3$ and $\varepsilon_{\mathrm{vol}} = 3\varepsilon_1 = 3\varepsilon_2 = 3\varepsilon_3 \neq 0$) is considered, as it allows the use of a single constitutive equation, contrary to oedometric compression ($T_1 \neq T_2 = T_3$ and $\varepsilon_1 \neq 0$, $\varepsilon_2 = \varepsilon_3$). For the compression test, the tangential stiffness is $K = \frac{\dot{p}}{\dot{\varepsilon}_{\mathrm{vol}}}$. The relationship of Ohde [90] and Janbu [57] can be expressed as follows[2]

$$K \propto \left(\frac{p}{p_{\mathrm{r}}}\right)^{\xi} \quad \text{or} \quad K = K_{\mathrm{r}} \left(\frac{p}{p_{\mathrm{r}}}\right)^{\xi} \quad , \tag{3.2}$$

p_{r} is a reference pressure (usually $1\,\mathrm{kPa}$ or the atmospheric pressure p_{at}), ξ and K_{r} are material parameters, which depend on the density of the soil. For clays ξ can be set to 1, for sands ξ is between 0.5 and 1.

From

$$K = \frac{\dot{p}}{\dot{\varepsilon}_{\mathrm{vol}}} = K_{\mathrm{r}} \left(\frac{p}{p_{\mathrm{r}}}\right)^{\xi} \tag{3.3}$$

it follows that

$$\dot{p} = K_{\mathrm{r}} \left(\frac{p}{p_{\mathrm{r}}}\right)^{\xi} \dot{\varepsilon}_{\mathrm{vol}} \quad . \tag{3.4}$$

In Fig. 3.4 results obtained with this differential equation are compared with laboratory tests of Hostun sand [22] and Weald clay [72] (from [91]).

The stiffness definition (3.2) can be explicitly used as stiffness term in Barodesy.

3.2.2   Asymptotic states

Soil is characterized by its asymptotic behaviour and the ability to "forget" its deformation history slowly. This ability is also known as "swept out of memory" effect.

Goldscheider [36] carried out numerous tests on sand with the True Triaxial Test device and was able to formulate two basic rules based on them:

---

[2] Originally Ohde [90] and Janbu [57] described herewith an oedometer test with $\varepsilon_2 = \varepsilon_3 = 0$ and $T_1 \neq T_3$, but it is also applicable to isotropic compression.

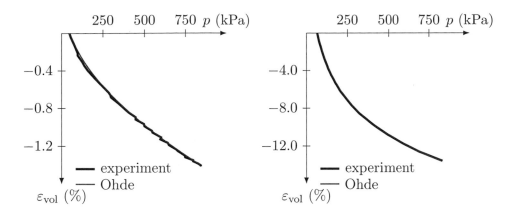

(a) Sand $K_r = 3175\,\text{kPa}$ and $\xi = 0.5$ (b) Clay $K_r = 19\,\text{kPa}$ and $\xi = 1$

Figure 3.4: Isotropic compression tests for Hostun sand [22] and Weald clay [72] compared with numerical results calculated with the compression law of Ohde

1. If a soil sample is deformed with a proportional strain path starting very close to the stress-free state, a proportional stress path is obtained

2. If a deformation with a proportional strain path starting from an admissible stress state, the stress path will asymptotically approach the proportional stress path corresponding to the proportional strain path starting from a stress-free state (following the first Goldscheider rule). Therefore, the soil "forgets" its original stress state with ongoing deformation

These rules form the basis of Barodesy and are experimentally proven for sand and clay [16, 36, 118, 119]. The stress paths of the experiments with Sydney sand Fig. 3.5 approach asymptotically proportional stress paths, also the corresponding proportional strain paths are shown. This indicates a correlation between proportional strain paths and proportional stress paths.

Mathematical definition of the first rule of Goldscheider

The objective stress rate $\mathring{\boldsymbol{T}}$ can be represented as

$$\mathring{\boldsymbol{T}} = \|\mathring{\boldsymbol{T}}\|\mathring{\boldsymbol{T}}^0 \quad , \tag{3.5}$$

where $\|\mathring{\boldsymbol{T}}\|$ is the norm of the objective stress rate and $\mathring{\boldsymbol{T}}^0$ is the normalized stress rate. The same applies analogously to the stress state

$$\boldsymbol{T} = \|\boldsymbol{T}\|\boldsymbol{T}^0 \quad . \tag{3.6}$$

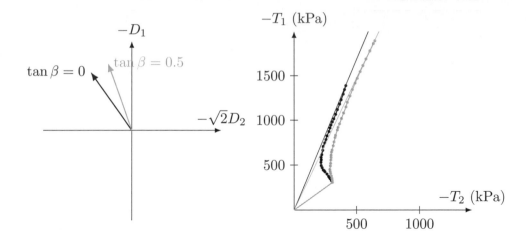

Figure 3.5: Triaxial test results with proportional strain paths, data from Chu and Lo [16]

We now consider a deformation rate $\boldsymbol{D}$, which causes the direction $\boldsymbol{R}^0$ for the proportional stress path, and a stress state $\boldsymbol{T}$ with the normalized stress tensor $\boldsymbol{T}^0$, which corresponds to the current deformation rate $\boldsymbol{D}$, i.e. $\boldsymbol{R}^0 = \boldsymbol{T}^0$. In this case, $\boldsymbol{T}^0$ of the stress state is also equal to the stress rate $\overset{\circ}{\boldsymbol{T}}{}^0$, i.e.

$$\boldsymbol{R}^0(\boldsymbol{D}) = \boldsymbol{T}^0 = \overset{\circ}{\boldsymbol{T}}{}^0 \quad . \tag{3.7}$$

This means that in the case of a deformation without a rotation (i.e. $\dot{\boldsymbol{T}}{=}\overset{\circ}{\boldsymbol{T}}$) the stress remains on the proportional stress path, e.g. for a step of the time integration with the explicit Euler method:

$$\boldsymbol{T}_{t+\Delta t} = \boldsymbol{T}_t + \dot{\boldsymbol{T}}_t \Delta t = \boldsymbol{T}_t^0 \|\boldsymbol{T}_t\| + \dot{\boldsymbol{T}}_t^0 \|\dot{\boldsymbol{T}}_t\| \Delta t = \boldsymbol{T}_t^0 \left(\|\boldsymbol{T}_t\| + \|\dot{\boldsymbol{T}}_t\| \Delta t \right)$$

R-function

From the first rule of Goldscheider it follows directly that for every possible proportional strain path, an associated proportional stress path must be defined. This is described in Barodesy with the so-called $\boldsymbol{R}$-function. The direction of a proportional strain path is determined by the stretching tensor $\boldsymbol{D}$, or the normalized stretching tensor $\boldsymbol{D}^0$. Arbitrary (volume reducing or volume increasing) proportional strain paths are possible. However, volume increasing strain paths are not limitless, as the void ratio cannot exceed a maximum value without loosing grain contacts completely. All proportional stress paths should remain within the permissible stress octant (with only compressive stresses). This is guaranteed by a function of the form

$$\boldsymbol{R} = -\exp\left(\alpha \boldsymbol{D}^0\right) \tag{3.8}$$

which in the case of a diagonal matrix can simply be written as

$$\begin{bmatrix} R_1 & 0 & 0 \\ 0 & R_2 & 0 \\ 0 & 0 & R_3 \end{bmatrix} = - \begin{bmatrix} \exp\left(\alpha D_1^0\right) & 0 & 0 \\ 0 & \exp\left(\alpha D_2^0\right) & 0 \\ 0 & 0 & \exp\left(\alpha D_3^0\right) \end{bmatrix} . \tag{3.9}$$

The function (3.8) is negative semidefinite, i.e. all eigenvalues are less than or equal to 0, due to the properties of the exponential function. This means that all stress paths are in the compressive stress octant. The scalar function α in (3.8) is fitted to the stress strain relationship.

Selected proportional strain and stress paths are shown in Fig. 3.6. The proportional strain path for isotropic compression (i) is described with $D_1 = D_2 = D_3 = -1$ what leads to $\operatorname{tr} \boldsymbol{D}^0 = \delta = -\sqrt{3}$ and $\psi_D = 0°$. The associated proportional stress path describes the isotropic stress axis ($\psi_T = 0°$). For axisymmetric isochoric compression (c) ($\operatorname{tr} \boldsymbol{D}^0 = \delta = 0$ and $\psi_D = 90°$) the proportional stress path is the critical state line in the stress space. A proportional strain path between (i) and (c) produces a proportional stress path between (i) and (c), e.g. the oedometric compression path (o) with $D_1 = -1$ and $D_2 = D_3 = 0$ ($\operatorname{tr} \boldsymbol{D}^0 = \delta = -1$ and $\psi_D = 54.74°$) results in the so-called K_0-path. However, also volume increasing proportional strain paths are possible. These paths produce mobilised friction angles φ_{mob} higher than the critical friction angle φ_{c} and do not allow an unlimited deformation. According to Gudehus [46] the theoretical maximum mobilised friction angle φ_{max} is caused by the proportional strain path (e) with $D_1 = 0$ and $D_2 = D_3 = 1$ ($\operatorname{tr} \boldsymbol{D}^0 = \delta = \sqrt{2}$ and path ($-$e) $\psi_D = 144.74°$). In this case, the mobilised friction angle is $\varphi_{\mathrm{mob}} \approx 90°$ ($\psi_T \approx 54.74°$). For higher ψ_D, ψ_T decreases again. For the isotropic extension ($-$i) ($D_1 = D_2 = D_3 = 1$, what leads to $\operatorname{tr} \boldsymbol{D}^0 = \delta = \sqrt{3}$ and $\psi_D = 180°$) the proportional stress path follows the hydrostatic axis ($\psi_T = 0°$) again.

All proportional stress paths with $\delta = $ const generates cones in the compressive stress octant of the principal stress space, see Fig. 3.7. The deformation $\delta = -1$ generates cone 1. For oedometric compression this is the so-called K_0-path, which is marked with o in Fig. 3.7. Isochoric deformations ($\delta = 0$) results in cone 2, the cone of the critical stress states. The critical state line is shown in red. Dilatant strain paths with $\delta = 0.5$ generate the cone 3, and the cone 4, which is degenerated into a pyramid, limits the permissible stresses on the compression range.

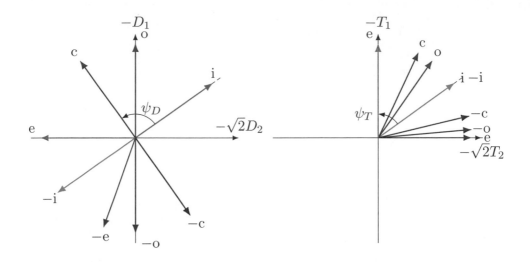

Figure 3.6: Proportional strain and stress paths for $\varphi_c = 30°$

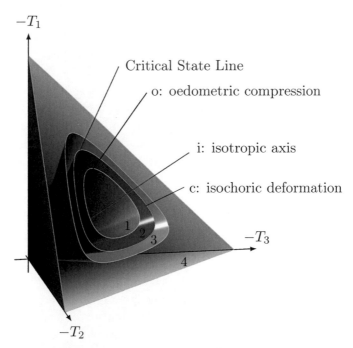

Figure 3.7: Proportional stress paths with $\operatorname{tr} \boldsymbol{D}^0 = \text{const}$ in the principal stress space (modified from Medicus *et al.* [78])

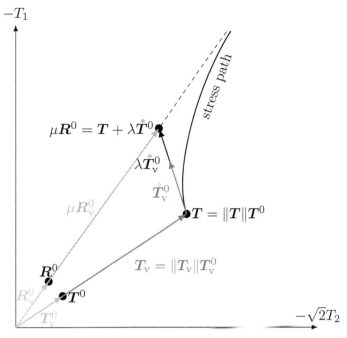

Figure 3.8: Visualisation of the asymptotic approach defined by (3.10)

Mathematical definition of the second rule of Goldscheider

The sample is deformed with a proportional strain path, starting from an arbitrary stress state $\boldsymbol{T}$. The stress path approaches a proportional stress path of the direction $\boldsymbol{R}^0$, visualised by the vector $\boldsymbol{R}_v^0$ in Fig. 3.8. The stress rate vector $\mathring{\boldsymbol{T}}_v^0$ is the tangent to the stress path and thus points in the direction of the proportional stress path. Mathematically, the asymptotic approach in the principal stress space can be described with

$$\|\boldsymbol{T}\|\boldsymbol{T}^0 + \lambda\mathring{\boldsymbol{T}}^0 = \mu\boldsymbol{R}^0 \quad , \tag{3.10}$$

which is visualised as vector addition in Fig. 3.8. Note that the vector sum $\boldsymbol{T}_v + \lambda\mathring{\boldsymbol{T}}_v^0$ points to the stress point defined by $\mu\boldsymbol{R}^0$, cf. Fig. 3.1c. Transformation of this equation leads to

$$\mathring{\boldsymbol{T}}^0 = \frac{\mu}{\lambda}\boldsymbol{R}^0 - \frac{\|\boldsymbol{T}\|}{\lambda}\boldsymbol{T}^0 \quad . \tag{3.11}$$

The scalar quantities μ and λ are functions of the void ratio and the stress level, since the direction and evolution of the approach depends at least on these two state variables. With $f = \dfrac{\mu}{\lambda}$ and $g = -\dfrac{\|\boldsymbol{T}\|}{\lambda}$ follows

$$\mathring{\boldsymbol{T}}^0 = f\boldsymbol{R}^0 + g\boldsymbol{T}^0 \quad . \tag{3.12}$$

Now we know the direction of the stress path $\mathring{T}^0$ and still need the value of the stress rate. This value follows from the generalisation of the one-dimensional consideration for the stress-dependant stiffness, cf. (3.2)

$$\|\mathring{T}\| = K_{\mathrm{r}} \left(\frac{\|T\|}{p_{\mathrm{r}}} \right)^{\xi} \|D\| = h(T)\|D\| \quad . \tag{3.13}$$

Note, instead of the mean stress p the generalized form $\|T\|$ is here used. Thus, from (3.12) and (3.13) follows with $\mathring{T} = \|\mathring{T}\|\mathring{T}^0$ the basic structure of Barodesy

$$\mathring{T} = h(f R^0 + g T^0)\|D\| \quad . \tag{3.14}$$

With ongoing proportional deformation, the stress state T asymptotically approaches the proportional stress path defined by R^0. After a large deformation $T^0 \simeq R^0$ holds true.

3.2.3   Critical state

The critical state in soil mechanics is characterized by vanishing stiffness ($\dot{p} = 0$ and $\dot{q} = 0$) and constant volume ($\dot{e} = 0$) for ongoing shearing and was first described by Roscoe *et al.* [96] as

$$\frac{\partial p}{\partial \varepsilon_q} = \frac{\partial q}{\partial \varepsilon_q} = \frac{\partial e}{\partial \varepsilon_q} = 0 \quad . \tag{3.15}$$

There is a relationship between the mean stress p, the deviatoric stress q and the void ratio e, which is described by the so-called critical state line. In the p-q diagram, the critical state line (CSL) is a straight line given by the equation

$$q = Mp \quad \text{with} \quad M = \frac{6 \sin \varphi_{\mathrm{c}}}{3 - \sin \varphi_{\mathrm{c}}} \tag{3.16}$$

for axisymmetric loading. This line is experimentally well documented, cf. Fig. 3.9.

For non-axisymmetric tests, the representation in the deviatoric plane is more suitable. Fig. 3.10 shows critical stress states of experiments with San Francisco Bay Mud ($\varphi_{\mathrm{c}} = 30.6°$) from Lade [67]. The samples were isotropic consolidated and sheared in conventional and True Triaxial devices. These results are compared with the Mohr-Coulomb and Matsuoka-Nakai [73] failure criteria. For axisymmetric loadings, the failure criteria are the same, for all other loadings they differ. Compared with the test results the Matsuoka-Nakai criterion agrees quite well.

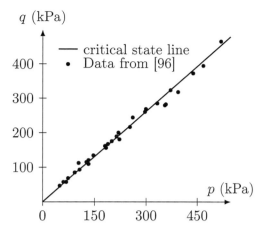

Figure 3.9: Critical state line in the *p-q* plane with test data from Roscoe *et al.* [96]

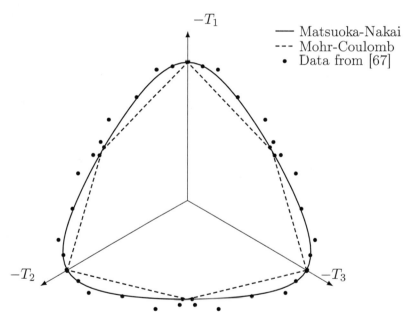

Figure 3.10: Critical stress states of San Francisco Bay Mud and failure criteria according to Matsuoka-Nakai [73] and Mohr-Coulomb (principal stresses are not sorted)

Barodesy does not use the Matsuoka-Nakai failure criterion. However, the $\boldsymbol{R}$-function delivers for isochoric deformations nearly identical results (cf. Fellin and Ostermann [33]).

The relation between stress and critical void ratio e_c is not that clear. According to Muir Wood [84] the critical state line is (at least for sand) a somewhat diffuse, but clearly pressure dependant zone. The critical state line in the pressure-void ratio diagram divides dense samples ($e < e_c$) from loose samples ($e > e_c$). There are many different formulations for the critical state line, both for clay and sand. Logarithmic (Mašín [74])

$$\ln(1 + e_c) = N - \lambda^* \ln \frac{2p}{\sigma^*} \qquad (3.17)$$

and semi-logarithmic approaches are common for clay and sand (Roscoe *et al.* [96], Gajo and Muir Wood [34])

$$e_c = e_\lambda - \lambda_e \ln \frac{p}{\sigma} \qquad . \qquad (3.18)$$

In these equations N or e_λ are the corresponding critical void ratios e_c to the reference stress level σ^* and σ, respectively. The inclination of the e_c-line in the $\ln e$-$\ln p$- and e-$\ln p$-plot is defined by the parameter λ^* and λ_e, respectively. For sand exponential and power approaches are also common (Bauer [4], Li and Wang [68])

$$e_c = e_{c0} \exp\left(-\left(\frac{3p}{h_s}\right)^{n_H}\right) \quad \text{or} \qquad (3.19)$$

$$e_c = e_{c0} - \lambda_e \left(\frac{p}{p_{at}}\right)^{\xi_e} \qquad . \qquad (3.20)$$

Here e_{c0} denotes the critical void ratio at zero stress, h_s is a stiffness parameter, p_{at} is the atmospheric pressure, n_H and ξ_e are curve fitting material parameters.

Fig. 3.11 shows the two approaches according to Mašín [74] and Roscoe *et al.* [96] for Weald clay in a logarithmic and a non-logarithmic representation, additionally the test data of [96] are also included. It can be clearly seen that both approaches deliver good results for the range of experiments and do not differ considerably from each other.

Various approaches for sand are shown in Fig. 3.12. Again, test results were added (Hostun Sand from Gajo and Muir Wood [34]). It can be seen that the various approaches differ more strongly from each other and that the test results scatter more than for clay.

The projection of the critical state line in the p-e plane is used in Barodesy to determine whether a specimen is loose or dense.

For axisymmetric stress states, the mean stress p, the deviatoric stress q and the void ratio e span a three-dimensional space in which the critical state line appears as curve (cf. Fig. 3.13). A specimen is only in the critical state if its

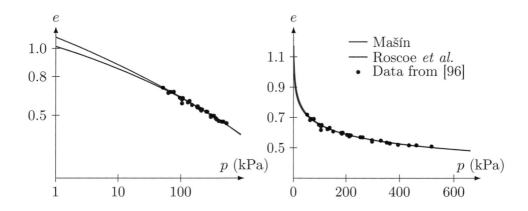

Figure 3.11: Different approaches for the critical state lines of Weald clay in the e-p-plane with results from [96].

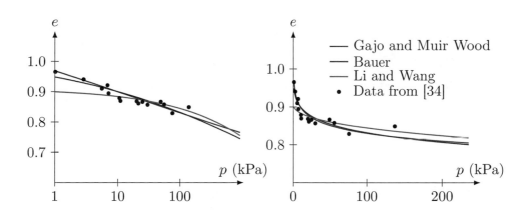

Figure 3.12: Different approaches for the critical state lines of Hostun sand in the e-p-plane with data from [34]

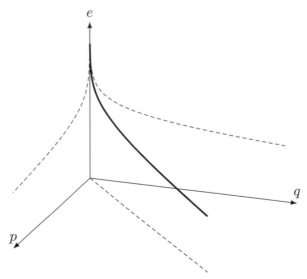

Figure 3.13: Critical state line in the three dimensional *p-q-e* space with its projections

state variables are on this space curve. It is not sufficient that the projection of the actual state is in the *p-q* plane or the *p-e* plane on the critical state line.

If the stress is in the critical state and the deformation is isochoric then $\boldsymbol{R}^0 = \boldsymbol{T}^0$ is valid. So (3.14) becomes

$$\mathring{\boldsymbol{T}} = h(f + g)\boldsymbol{R}^0\|\boldsymbol{D}\| \quad . \tag{3.21}$$

For the critical state $\mathring{\boldsymbol{T}} = \boldsymbol{0}$, and from (3.21) it follows that in critical state $f + g = 0$ is required. From this, conditions for the formulation of f and g can be derived. In particular, the critical state line can be directly incorporated.

3.2.4   Stress-dilatancy relations

The mobilised friction angle is associated with the dilatancy (Rowe [97], Taylor [112]). If the soil is dilatant, the peak friction angle φ_p is higher than the critical friction angle φ_c.

Chu and Lo [16], like Goldscheider [38], have deformed soil samples with proportional strain paths starting from an isotropic stress state, i.e. the dilatancy $\tan\beta$ is constant throughout the entire experiment. Chu and Lo observed that the achievable stress ratios η (in the asymptotic state) are related to the dilatancy of the proportional strain paths $\tan\beta$. Based on their test results,

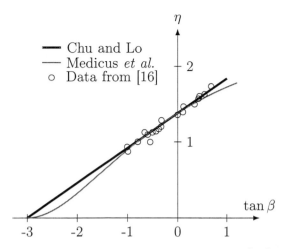

Figure 3.14: Comparison of the relation of Chu and Lo [16] and Barodesy with experimental data of Sydney Sand [16] ($\varphi_c = 34.41°$)

Chu and Lo [16] introduced the following empirical relationship between the mobilised friction angles and proportional strain paths

$$\eta = \frac{q}{p} = \frac{M}{3}\tan\beta + M \quad .\tag{3.22}$$

The higher the dilatancy $\tan\beta$, the higher the achievable stress ratio η. If the dilatancy measure is $-3 < \tan\beta < 0$, the mobilised friction angle φ_{mob} is lower than the critical friction angle φ_c, cf. state of the oedometric compression ($\tan\beta = -1$ and $\varphi_{mob} < \varphi_c$). For an isotropic compression ($\tan\beta = -3$) the mobilised friction angle $\varphi_{mob} = 0$ and $\eta = 0$, since $q = 0$. Fig. 3.14 shows a comparison of Barodesy in the version of Medicus *et al.* [79] with the relationship according to Chu and Lo [16]. In the range of the experimental results ($-1 < \tan\beta < 0.8$) Barodesy gives very similar results as the relationship according to Chu and Lo (cf. Fig 3.14).

Also at peak states, there is a correlation between dilatancy and peak friction angle φ_p. The higher the dilatancy $\tan\beta$ in the peak, the higher the peak friction angle φ_p, see Fig. 3.15, which shows triaxial tests of Hostun Sand by Desrues *et al.* [22]. The peak states of these consolidated drained experiments can also be estimated with the relationship according to Chu and Lo.

3.2.5 Pyknotropy and Barotropy

The denser a soil specimen is, the higher its peak strength is (with the same initial stress) and thus its peak friction angle φ_p. The influence of the void

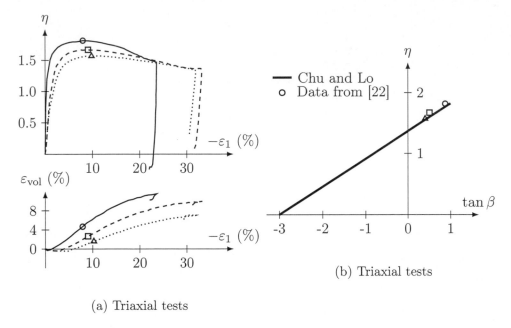

(a) Triaxial tests

(b) Triaxial tests

Figure 3.15: Peak state of consolidated drained triaxial test on Hostun sand [22]

ratio is called pyknotropy. Fig. 3.16 shows simulation results with Barodesy for clay [76] for illustration. Two drained triaxial tests are simulated. In both simulations the initial consolidation pressure is $p_{ini} = 200\,\mathrm{kPa}$. The simulations differ just in their initial void ratios (dense $e_{ini} = 0.8$ and loose $e_{ini} = 1.2$). The connection between stress ratio η and dilatancy δ also exists here: the dense sample ($e_{ini} < e_c$) must loosen in order to reach the critical state, the loose sample ($e_{ini} > e_c$) must compact in order to reach the critical state, cf. Fig. 3.16. The strength q_{max} and the peak friction angle φ_p are higher for the dense sample than for the loose one.

If consolidated drained tests are simulated at different stress levels (initial stress $p_{ini} = 100\,\mathrm{kPa}$ and $p_{ini} = 300\,\mathrm{kPa}$) with the same initial void ratio ($e_{ini} = 1$), then the results shown in Fig. 3.17 are obtained. Although both soil samples have the same initial void ratio, the sample with the initial stress $p_{ini} = 100\,\mathrm{kPa}$ is "dense" by definition ($e_{ini} < e_c$), and the sample with the initial stress $p_{ini} = 300\,\mathrm{kPa}$ by definition "loose" ($e_{ini} > e_c$). Thus, the terms pyknotropy and the barotropy (influence of the stress level) are inseparably linked. Here too, the following applies: the dense sample must loosen in order to reach the critical state, the loose sample must compact in order to get to the critical state. The sample with the higher stress level ($p_{ini} = 300\,\mathrm{kPa}$) achieves a higher maximum deviatoric stress q_{max} than the sample with the lower stress level ($p_{ini} = 100\,\mathrm{kPa}$), see Fig. 3.17. If the deviatoric stress q is

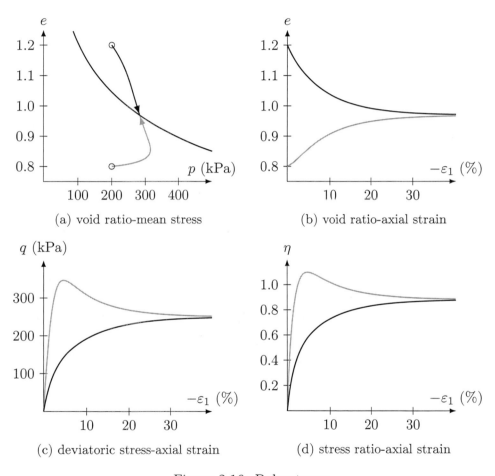

(a) void ratio-mean stress

(b) void ratio-axial strain

(c) deviatoric stress-axial strain

(d) stress ratio-axial strain

Figure 3.16: Pyknotropy

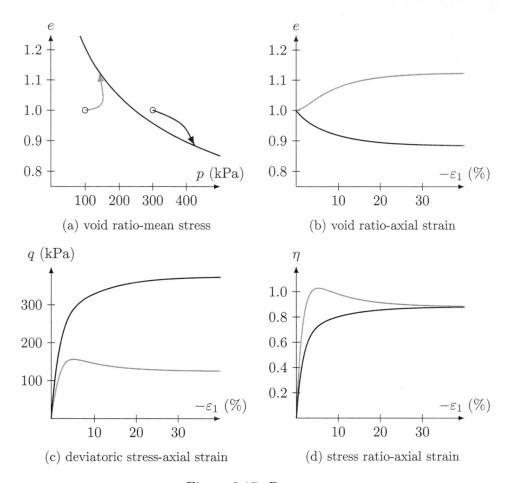

(a) void ratio-mean stress

(b) void ratio-axial strain

(c) deviatoric stress-axial strain

(d) stress ratio-axial strain

Figure 3.17: Barotropy

now normalized by the mean stress p, the maximum stress ratio η and also the peak friction angle φ_{p} for the dense sample (small stress level) are higher than for the loose sample (high stress level).

Not only in drained test the influence of the void ratio can be seen. In Fig. 3.18 results of simulations with a dilatant proportional strain path ($D_1 = 0$ and $D_2 = D_3 > 0$) are shown. The simulation start at the same isotropic stress state and differ only in the initial void ratio (dense, critical and loose). Fig. 3.18a shows the evolution of the deviatoric stress q. Also here the dense sample gains a higher peak strength than the one with the critical void ratio at start, and the loose sample has the lowest peak strength. However, all three samples reach the same stress ratio η and mobilised friction angle φ_{mob} at the end of the tests (cf. Fig. 3.18b). This stress ratio is the attractor for this specific proportional strain path. From these diagrams it can be seen that

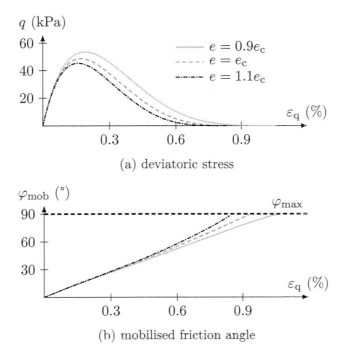

(a) deviatoric stress

(b) mobilised friction angle

Figure 3.18: Evolution of stresses for a proportional strain path and various void ratios

for all three void ratios the maximum of the mobilised friction angles is the same. Although the deviatoric stresses q are quite different.

In Fig. 3.19 the according stress paths in the Rendulic plane for the latter tests are shown. Starting from the isotropic stress state all stress paths heading to the origin, but they are not following the same path. It can be also observed, that the stress paths exceed the cone of critcal stresses for all densities, i.e. the mobilised friction angle φ_{mob} is higher then the critical friction angle φ_c. The reason for this is the attractor for the dilatant strain path R_v^0.

3.3   Barodesy for clay

The basic structure of Barodesy is

$$\mathring{\boldsymbol{T}} = h(f\boldsymbol{R}^0 + g\boldsymbol{T}^0)\|\boldsymbol{D}\| \tag{3.23}$$

where

$$\boldsymbol{R} = -\exp(\alpha\boldsymbol{D}^0) \tag{3.24}$$

is used to calculate the proportional stress paths. This structure is the same for sand and clay. However, the used scalar functions h, f, g and α differ for

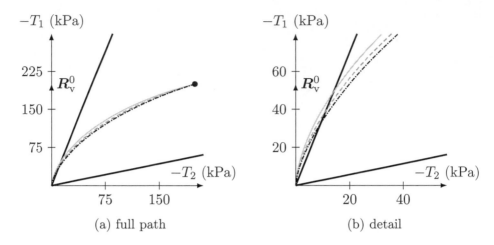

(a) full path   (b) detail

Figure 3.19: Stress paths for a proportional strain path and various void ratios

both materials. In this section, the functions and the calibration for clay as presented in Medicus and Fellin [76] will be described.

The scalar function for α in $\boldsymbol{R}$ is

$$\alpha = \frac{\sqrt{2}\ln K_\delta}{\sqrt{3 - \delta^2}} \quad , \tag{3.25}$$

with

$$K_\delta = 1 - \frac{1}{1 + c_1(m_B - c_2)^2} \quad \text{and} \quad m_B = \frac{-3\delta}{\sqrt{6 - 2\delta^2}} \quad .$$

With (3.25) Barodesy can map the following stress-dilatancy relations

- a mobilised friction angle $\varphi_{\text{mob}} = 0°$ for hydrostatic compression

- for oedometric compression the $\boldsymbol{R}$-function follows the simplified relation of Jáky [56] ($K_0 = 1 - \sin\varphi_c$)

- for critical states the mobilised friction angle is equal to the critical friction angle $\varphi_{\text{mob}} = \varphi_c$

- for proportional strain paths as described in Chu and Lo [16] the $\boldsymbol{R}$-function provides similar results.

- all proportional stress paths are limited to the compression octant, no tensile stresses can occur.

The stiffness term h reads

$$h = c_3\|\boldsymbol{T}\|^{c_4} \quad . \tag{3.26}$$

The scalar functions f and g contain the current and the critical void ratio (and thus the actual stress level). The functions are

$$f = c_6 b\delta - \frac{1}{2} \quad \text{and} \tag{3.27}$$

$$g = (1 - c_6)b\delta + \left(\frac{1+e}{1+e_c}\right)^{c_5} - \frac{1}{2} \quad . \tag{3.28}$$

The logarithmic relation is used for the critical void ratio

$$e_c = \exp\left(N - \lambda^* \ln \frac{2p}{\sigma^*}\right) - 1 \tag{3.29}$$

The other equations required are

$$b = \frac{1}{c_3\Lambda} + \frac{1}{\sqrt{3}}2^{c_5\lambda^*} - \frac{1}{\sqrt{3}} \quad \text{and} \tag{3.30}$$

$$\Lambda = -\frac{\lambda^* - \kappa^*}{2\sqrt{3}}\delta + \frac{\lambda^* + \kappa^*}{2} \quad . \tag{3.31}$$

The term $\left(\frac{1+e}{1+e_c}\right)$ in (3.28) expresses the distance of the current to the critical void ratio. This determines the current overconsolidation via a relative density. The peak states and stiffness, which depend on stress and void ratio, can thus be realistically estimated. If the void ratio is equal to the critical void ratio $e = e_c$ in an isochoric deformation then $f + g = 0$. Thus, the stress rate vanishes in the critical state, i.e. $\mathring{\boldsymbol{T}} = \boldsymbol{0}$.

Equations (3.30) and (3.31) ensure, among other things, that normal consolidated[3] soil under hydrostatic compression follows the so-called normal compression line; λ^* is the relevant stiffness parameter. For hydrostatic unloading, κ^* expresses the stiffness, cf. Fig. 3.20. Parameters c_1 to c_6 are constants that can be determined from the four material parameters φ_c, λ^*, κ^* and N with

---

[3] in the case of sand the loosest possible packing for isotropic conditions

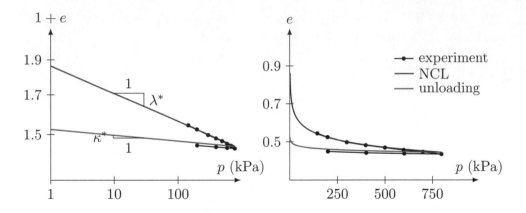

Figure 3.20: Normal compression line and unloading

the following equations

$$c_1 = \frac{1 - \sin \varphi_c}{2c_2^2 \sin \varphi_c} \quad ,$$

$$c_2 = -\frac{3\sqrt{2} + 3}{2} \approx -3.6213 \quad ,$$

$$c_3 = \frac{-\frac{\sqrt{3}}{\lambda^*} + \frac{\sqrt{3}}{\kappa^*}}{2^{c_5\lambda^*} + 0.002^{c_5\lambda^*} - 2} \quad ,$$

$$c_4 = 1 \quad ,$$

$$c_5 = \frac{1 + \sin \varphi_c}{1 - \sin \varphi_c} \quad \text{and}$$

$$c_6 = \frac{1}{2\left(\frac{-\sqrt{3}}{c_3\kappa^*} + 2^{c_5\lambda^*} - 1\right)} \quad .$$

For further details on the mathematical formulation, please see Medicus [75].

3.4 Barodesy for sand

For sand, the basic structure of Barodesy is the same as for clay. The two versions just differ in the scalar functions. Here, the version of Kolymbas [64] is described.

The scalar function for α in $\boldsymbol{R}$ is

$$\alpha = c_1 \exp(c_2\delta) \quad . \tag{3.32}$$

The scalar functions f, g and h are

$$f = \delta + c_3 e_c \quad , \tag{3.33}$$

$$g = -c_3 e \quad \text{and} \tag{3.34}$$

$$h = -\frac{c_4 + c_5 \|\boldsymbol{T}\|}{e - e_{\min}} \quad . \tag{3.35}$$

The equation for the critical void ratio can be derived from the functions of Barodesy and reads

$$e_c = \frac{e_{\min} + B}{1 - B} \tag{3.36}$$

with

$$B = \frac{e_{c0} - e_{\min}}{e_{c0} + 1} \left(\frac{c_4 + c_5 \|\boldsymbol{T}\|}{c_4} \right)^{-(1 + e_{\min})/c_5} \quad . \tag{3.37}$$

The constant c_1 can be derived from the critical friction angle φ_c

$$c_1 = \sqrt{\frac{2}{3}} \ln \left(\frac{1 - \sin \varphi_c}{1 + \sin \varphi_c} \right) \quad . \tag{3.38}$$

The further constants (c_2, c_3, c_4, c_5, e_{c0} and $e_{\min}$) can be calibrated on the basis of triaxial and oedometric tests. For the detailed procedure see Kolymbas [64].

Chapter 4

Further developments for Barodesy

4.1 New formulation for the R-function

The R-function introduced by Medicus *et al.* [79] delivers quite good results. However, a division by 0 is possible and causes numerical problems and the formulation is not very elegant, since the scalar quantity α is a quite complex function of the direction of the stretching D^0. In contrary, the R-function introduced in Kolymbas [64] is quite simple and has no poles, but the results are not as good (cf. Fig. 4.1 and 4.2). This function is also not able to predict stresses in peak states (cf. Fig. 4.1) and so the maximum mobilised friction angle is not much higher than the critical one.

The aim of a new formulation is to describe the stress dilatancy behaviour of soil with a quite simple function, which has no poles. In the following, only rectilinear deformations and asymptotic states are considered. Therefore, the first rule of Goldscheider is applicable and $R^0 = T^0$ holds true.

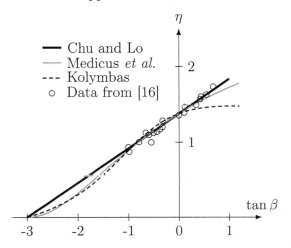

Figure 4.1: Comparison of various R-functions with the relation from Chu and Lo [16] (for $\varphi_c = 30°$)

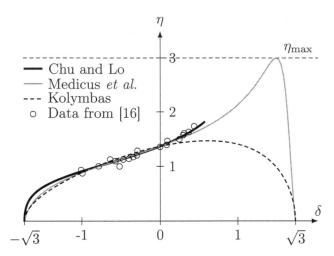

Figure 4.2: Comparison of results of the $\boldsymbol{R}$-function with different formulations and the relation from Chu and Lo [16] plotted vs. δ (for $\varphi_\mathrm{c} = 30°$)

4.1.1   The function of Chu and Lo

It appears promising to use the relation of Chu and Lo for formulation of α. The original function is

$$\eta = \frac{q}{p} = M + \frac{M}{3}\tan\beta \quad . \tag{4.1}$$

This function has the advantage that the stress-dilatancy relations for the critical states and the peak states is in accordance with the one by Chu and Lo [16]. The relation for oedometric compression is also the same as in [16], which is slightly different from the formulation postulated by Jáky [56] ($K_0 = 0.9(1 - \sin\varphi)$ usually simplified to $K_0 = 1 - \sin\varphi$), but also within the range of experimental accuracy.

With the definition of $\tan\beta$ and the assumption of axisymmetric deformation, we can conclude that

$$D_3 = -D_1\frac{1 + \tan\beta}{2} \tag{4.2}$$

and hence the norm is

$$\|\boldsymbol{D}\| = \sqrt{\mathrm{tr}\,\boldsymbol{D}^2} = \sqrt{D_1^2 + 2D_3^2} = |D_1|\sqrt{1 + \frac{(1 + \tan\beta)^2}{2}} \quad . \tag{4.3}$$

With these expressions the $\boldsymbol{R}$-function

$$\boldsymbol{R} = -\exp\left(\alpha\boldsymbol{D}^0\right)$$

can be written as

$$
\boldsymbol{R} = -\exp\left(\frac{-\alpha}{2\sqrt{1+\frac{(1+\tan\beta)^2}{2}}}\begin{bmatrix} -2 & 0 & 0 \\ 0 & 1+\tan\beta & 0 \\ 0 & 0 & 1+\tan\beta \end{bmatrix}\right) . \tag{4.4}
$$

To substitute η in (4.1) the deviatoric stress and the mean stress are calculated. Therefore the relation between R_1 and R_3 is defined as

$$
K_T = \frac{T_3}{T_1} = \frac{R_3}{R_1} = \frac{-\exp\left(\frac{-\alpha(1+\tan\beta)}{2\sqrt{1+\frac{(1+\tan\beta)^2}{2}}}\right)}{-\exp\left(\frac{2\alpha}{2\sqrt{1+\frac{(1+\tan\beta)^2}{2}}}\right)} = \exp\left(\frac{-\alpha(3+\tan\beta)}{2\sqrt{1+\frac{(1+\tan\beta)^2}{2}}}\right)
$$

$$\tag{4.5}$$

and the relation between deviatoric stress and mean stress can be rewritten as

$$
\eta = \frac{3(T_1 - T_3)}{T_1 + 2T_3} = \frac{3(1 - K_T)}{1 + 2K_T} . \tag{4.6}
$$

Now the equations (4.1) and (4.6) can be combined and solved for α, then

$$
\alpha = -\frac{\sqrt{2\tan^2\beta + 4\tan\beta + 6}}{3 + \tan\beta} \ln\left(-\frac{2M\tan\beta + 6M + 9}{M\tan\beta + 3M - 9}\right) . \tag{4.7}
$$

This results in a line which is identical with the relation postulated by Chu and Lo [16] between

$$
-3 < \tan\beta < \frac{9}{M} - 3 \tag{4.8}
$$

(cf. Fig. 4.3). The function α is also defined in a wider range, in particular between

$$
-\frac{9}{2M} - 3 < \tan\beta < \frac{9}{M} - 3 . \tag{4.9}
$$

For $\tan\beta$ smaller than $\frac{9}{2M} - 3$ and $\tan\beta$ higher than $\frac{9}{M} - 3$ the term in the logarithm gets smaller than zero, where the logarithm is not defined. In the range $\tan\beta$ smaller than -1 there are no experimental results, since these strain paths are strongly dilatant. For the strain paths in the Rendulic plane, this means that in the range of

$$
n\cdot\pi - 0.6155 - \arctan\frac{\sqrt{2}M}{9 - 2M} < \psi_D < n\cdot\pi - 0.6155 + \arctan\frac{\sqrt{8}M}{9 + 4M} \quad n \in \mathbb{N}. \tag{4.10}
$$

the α-function is not defined (cf. Fig. 4.4).

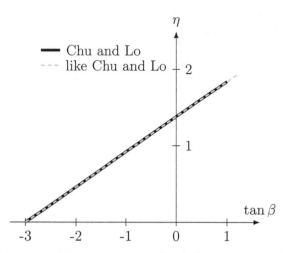

Figure 4.3: Comparison of the relation from Chu and Lo [16] and results of the $\boldsymbol{R}$-function with a formulation for α derived from [16] (for $\varphi_c = 30°$)

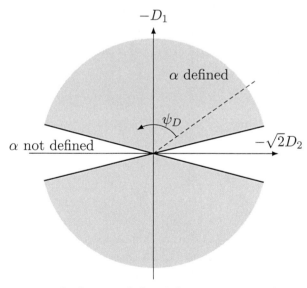

Figure 4.4: Areas in which α is defined for proportional strain paths in the Rendulic plane are shown in grey (for $\varphi_c = 30°$)

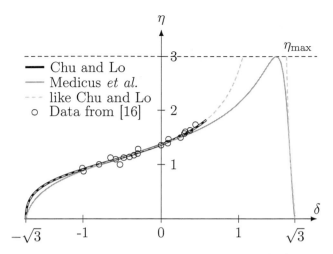

Figure 4.5: Comparison of relation from Chu and Lo [16] with results of the **R**-function with the relation derived from Chu and Lo vs. δ (for $\varphi_c = 30°$). Also the results of Medicus *et al.* [79] are shown.

This leads also to problems for the **R** function. In Fig. 4.5, it can be seen that for a critical state friction angle $\varphi_c = 30°$ no stress ratio η is defined in the range between $1.12 < \delta < 1.61$.

This function is not simpler than the function proposed by Medicus *et al.* [79] and has the disadvantage that it is not defined for all possible proportional strain paths. This makes it necessary to find another function for α.

4.1.2 New **R**-function

The function should be able to reproduce correct stress-dilatancy relation for critical state, oedometric compression and peaks. Additionally the function should never return $\boldsymbol{R} = \boldsymbol{0}$ (which causes numerical problems in the Barodesy since $\boldsymbol{0}/\boldsymbol{0}$ is not defined, and $\boldsymbol{R}^0$ has no direction) and the entire octant with negative stresses should be a possible solution.

To fulfil the additional conditions the scalar quantity α should be an element of $\mathbb{R}^-$. A possible function for α is

$$\alpha(\delta) = \alpha_{\min} + c_3 \frac{|\delta - x|^{c_2}}{(1 + |\delta - x|)^{c_1}} \tag{4.11}$$

here x is the dilatancy where the maximum mobilised friction angle occur. This function is able to map three different given strain states (oedometer, critical and peak states) to the corresponding stress states. Furthermore, $\alpha \in \mathbb{R}^-$ is ensured for $\alpha_{\min} \in \mathbb{R}^-$ and $c_3 \geq 0$.

It is possible to calculate the required value of the α-function for a given stress-ratio. In such cases the relation between T_1 and T_3 is known and hence also the relation between R_1 and R_3 for proportional paths. This leads to

$$K_T = \frac{R_3}{R_1} = \frac{-\exp\left(\alpha D_3^0\right)}{-\exp\left(\alpha D_1^0\right)} = \exp\left(\alpha\left(D_3^0 - D_1^0\right)\right) \tag{4.12}$$

from which α can be calculated to

$$\alpha = \frac{\ln K_T}{D_3^0 - D_1^0} \quad . \tag{4.13}$$

We will use the mobilised friction angle φ_{mob} for calibration. We have to use $K_T = \dfrac{R_3}{R_1}$ because the principal values of the stress tensor are sorted. Thus (4.13) changes to

$$\alpha = \frac{\ln K_T}{|D_3^0 - D_1^0|} \quad . \tag{4.14}$$

The function (4.11) has its minimum at $\delta = x$ with the value α_{min}. According to Gudehus [46] the maximum friction angle is attained for the two proportional stress paths along the two positive axes of the Rendulic plane

$$\boldsymbol{D}^0 = \begin{bmatrix} 1 & 0 & 0 \\ 0 & 0 & 0 \\ 0 & 0 & 0 \end{bmatrix} \quad \text{and} \quad \boldsymbol{D}^0 = \begin{bmatrix} 0 & 0 & 0 \\ 0 & \sqrt{1/2} & 0 \\ 0 & 0 & \sqrt{1/2} \end{bmatrix} \quad .$$

Therefore, the variable x should be chosen to 1 or $\sqrt{2}$. Note, an $\boldsymbol{R}$-function employing (4.11) is able to provide the maximum friction angle only for one of these dilatancies. This means that the maximal mobilised friction angle is reached either at $\delta = \sqrt{2}$ or $\delta = 1$. If this maximum mobilised friction angle φ_{max} should be 90° then K_T would be zero and $\ln K_T$ cannot be evaluated in (4.14). Therefore, a slightly smaller maximum mobilised friction angle φ_{max} has to be chosen to handle this situation numerically. Experiments on overconsolidated Dresden clay show a maximum mobilised friction angle of $\varphi_{\mathrm{max}} \approx 86°$ (Bergholz [8] and Bergholz and Herle [9]).

$\boldsymbol{\alpha_{\mathrm{min}}}$ **for** $x = 1 = \delta$**:** The value of α_{min} can be calculated with equation (4.14) and the relation between K_T and φ_{mob} as ($|D_3^0 - D_1^0| = |-1|$)

$$\alpha_{\mathrm{min}} = \ln\left(\frac{1 - \sin\varphi_{\mathrm{max}}}{1 + \sin\varphi_{\mathrm{max}}}\right) \tag{4.15}$$

and it can be seen as an internal variable, which is the same for every type of soil.

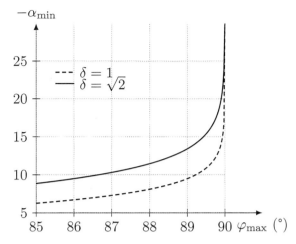

Figure 4.6: $\alpha_{\min}$ in dependence of the maximum mobilised friction angle

Table 4.1: Values for the calibration of the constants in the α-function

state	δ	K_T	$|D_3^0 - D_1^0|$	α
oedometric	-1	$K_0 = 1 - \sin\varphi_c$	1	α_0
critical	0	$K_c = \frac{1-\sin\varphi_c}{1+\sin\varphi_c}$	$\sqrt{\frac{3}{2}}$	α_c
peak	$\frac{1}{\sqrt{3}}$	$K_p = \frac{9-11\sin\varphi_c}{9+13\sin\varphi_c}$	1	α_p

$\alpha_{\min}$ **for** $x = \sqrt{2} = \delta$: As like as for $x = 1$ $\alpha_{\min}$ can be calculated with equation (4.14). This time the denominator is not 1, it is $|\sqrt{1/2}|$ and

$$\alpha_{\min} = \sqrt{2}\ln\left(\frac{1 - \sin\varphi_{\max}}{1 + \sin\varphi_{\max}}\right) \quad . \tag{4.16}$$

In Fig. 4.6, the connection between the maximum mobilised friction angle $\varphi_{\max}$ and $\alpha_{\min}$ is shown for both cases. In the following $\alpha_{\min}$ will be set to -30, where the two lines virtually coincides.[1]

The values from Tab. 4.1 for critical (c), oedometric (o) and peak states (p) are used in equation (4.11) to determine the constants c_1, c_2 and c_3. This

---

[1] In numbers $\varphi_{\max} = 89.997°$ for $x = \sqrt{2}$ and $\varphi_{\max} = 89.99996°$ for $x = 1$.

results in the following system of equations

$$\alpha_0 = \alpha_{\min} + c_3 \frac{|-1-x|^{c_2}}{(1+|-1-x|)^{c_1}}$$

$$\alpha_c = \alpha_{\min} + c_3 \frac{|0-x|^{c_2}}{(1+|0-x|)^{c_1}}$$

$$\alpha_p = \alpha_{\min} + c_3 \frac{\left|\sqrt{\frac{1}{3}}-x\right|^{c_2}}{\left(1+\left|\sqrt{\frac{1}{3}}-x\right|\right)^{c_1}} \quad .$$

The solutions of this system is

$$c_1 = \frac{\ln\left(\frac{\alpha_p-\alpha_{\min}}{\alpha_c-\alpha_{\min}}\right)\ln\left(\frac{1+x}{x}\right) - \ln\left(\frac{\alpha_0-\alpha_{\min}}{\alpha_c-\alpha_{\min}}\right)\ln\left(\frac{3x-\sqrt{3}}{3x}\right)}{\ln\left(\frac{1+x}{x}\right)\ln\left(\frac{3+3x}{3+3x-\sqrt{3}}\right) - \ln\left(\frac{1+x}{2+x}\right)\ln\left(\frac{3x-\sqrt{3}}{3x}\right)} \quad , \tag{4.17}$$

$$c_2 = \frac{\ln\left(\frac{\alpha_0-\alpha_{\min}}{\alpha_c-\alpha_{\min}}\right) - c_1\ln\left(\frac{1+x}{2+x}\right)}{\ln\left(\frac{1+x}{x}\right)} \quad \text{and} \tag{4.18}$$

$$c_3 = (\alpha_c - \alpha_{\min})\frac{(1+x)^{c_1}}{x^{c_2}} \quad . \tag{4.19}$$

For $x = 1$ the constants are then

$$c_1 = \frac{\ln 2 \ln\left(\frac{\alpha_p+30}{\alpha_c+30}\right) - \ln\left(\frac{\alpha_0+30}{\alpha_c+30}\right)\ln\left(\frac{3-\sqrt{3}}{3}\right)}{\ln 2 \ln\left(\frac{6}{6-\sqrt{3}}\right) - \ln\frac{2}{3}\ln\left(\frac{3-\sqrt{3}}{3}\right)} =$$

$$= -6.1293\ln\left(\frac{\alpha_p+30}{\alpha_c+30}\right) - 7.6155\ln\left(\frac{\alpha_0+30}{\alpha_c+30}\right) \quad ,$$

$$c_2 = \frac{\ln\left(\frac{\alpha_0+30}{\alpha_c+30}\right) - c_1\ln\left(\frac{2}{3}\right)}{\ln 2} = \quad \text{and}$$

$$c_3 = (\alpha_c+30)2^{c_1}$$

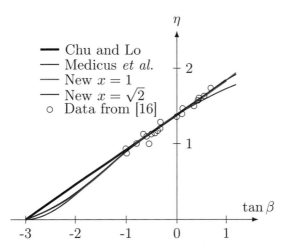

Figure 4.7: Comparison of relation from Chu and Lo [16] with results of the new **R**-function and the **R**-function from Medicus *et al.* [79]

and for $x = \sqrt{2}$ the constants are

$$c_1 = \frac{\ln\left(\frac{\alpha_p \mid 30}{\alpha_c+30}\right)\ln\left(\frac{1+\sqrt{2}}{\sqrt{2}}\right) - \ln\left(\frac{\alpha_0+30}{\alpha_c+30}\right)\ln\left(\frac{3\sqrt{2}-\sqrt{3}}{3\sqrt{2}}\right)}{\ln\left(\frac{1+\sqrt{2}}{\sqrt{2}}\right)\ln\left(\frac{3+3\sqrt{2}}{3+3\sqrt{2}-\sqrt{3}}\right) - \ln\left(\frac{1+\sqrt{2}}{2+\sqrt{2}}\right)\ln\left(\frac{3\sqrt{2}-\sqrt{3}}{3\sqrt{2}}\right)} =$$

$$= -14.9940\ln\left(\frac{\alpha_p+30}{\alpha_c+30}\right) - 14.6975\ln\left(\frac{\alpha_0+30}{\alpha_c+30}\right) \quad ,$$

$$c_2 = \frac{\ln\left(\frac{\alpha_0+30}{\alpha_c+30}\right) - c_1\ln\left(\frac{1+\sqrt{2}}{2+\sqrt{2}}\right)}{\ln\left(\frac{1+\sqrt{2}}{\sqrt{2}}\right)} = \frac{\ln\left(\frac{\alpha_0+30}{\alpha_c+30}\right) + 0.3466c_1}{0.5348} \quad \text{and}$$

$$c_3 = (\alpha_c + 30)\frac{(1+\sqrt{2})^{c_1}}{\sqrt{2}^{c_2}} \quad .$$

In Fig. 4.7 the new functions are compared with the relation from Chu and Lo [16] and Medicus *et al.* [79]. The new function is for both cases nearly the same for the shown range. The new function is also quite similar to the version from Medicus *et al.* [79], but produces higher peak strengths for dilatant strain paths and higher critical friction angles. For these areas the function is in good agreement with the relation from Chu and Lo [16].

From Fig. 4.8 and 4.9 it can be seen that the maximum mobilised friction angle φ_{max} is reached at $\delta = 1$ or $\delta = \sqrt{2}$ as desired. In these plots a quite different outcome for the two different cases with $x = 1$ and $x = \sqrt{2}$ can be seen. The differences between the proposed function by Medicus *et al.* [79] and new function with $x = 1$ are also obvious. However, the new function with $x = \sqrt{2}$ is similar to the version of Medicus *et al.* [79], but with higher

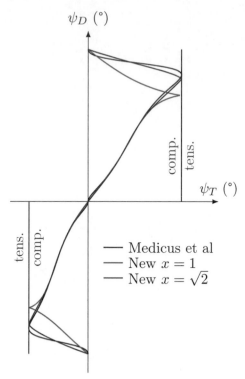

Figure 4.8: ψ_T vs. ψ_D is plotted for proportional strain paths (axes are not equal scaled), all resulting proportional stress paths are in the compressive quadrant ($-35.3° \leq \psi_T \leq 54.7°$)

peak friction angles for moderately high dilatancies (cf. Fig. 4.9).

Due to the quite similar results compared to the **R**-function from Medicus *et al.* [79] it is recommended to use (4.11) with $x = \sqrt{2}$.

4.1.3   Numerical experiments

In this section the numerical results with new **R**-function are compared with the version Medicus *et al.* [79] and laboratory tests. Therefore, Barodesy in the version of Medicus and Fellin [76] is used.

Triaxial tests

Experimental data of triaxial tests from Parry [91] (used in [72]) are used for comparison. In Fig. 4.10 simulations with the Barodesy version published in [76] and this version with a replaced **R**-function according to (4.11) with

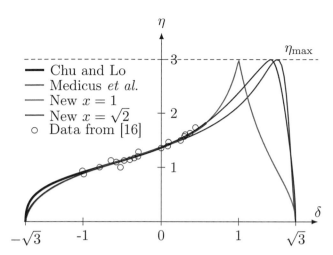

Figure 4.9: Comparison of relation from Chu and Lo [16] with results of the new **R**-function vs. δ. Also the result of Medicus *et al.* [79] are shown.

$x = \sqrt{2}$ are shown. It is obvious that there are no significant differences in the model predictions.

Oedometric tests

An oedometer test with London Clay from Mašín [71] is simulated with the two **R**-functions. Also in this test, the differences are pretty small and occur only in the unloading stress path, Fig. 4.11. This results from the fact that the two **R**-functions return the same values for oedometric compression, but not the same for oedometric extension ($\delta = 1$).

Proportional strain paths

Proportional strain paths are simulated to compare the new **R**-function with the one proposed by Medicus *et al.* [79]. For this purpose, a time integration of the constitutive model is performed with various stretchings **D** (Fig. 4.12a-left) starting at an initial stress state **T**$_0$.

In Fig. 4.12a, simulations starting from an isotropic stress state are shown. Results from the new **R**-function and the formulation [79] are shown in coloured lines and dashed black lines, respectively. In a wide range of ψ_D the results are the same. Only for proportional stress paths near the isotropic compression small differences occur.

The results for an initial stress state **T**$_0$ near the critical one in Fig. 4.12b are very similar to the results in Fig. 4.12a. Again, only the stress paths next to

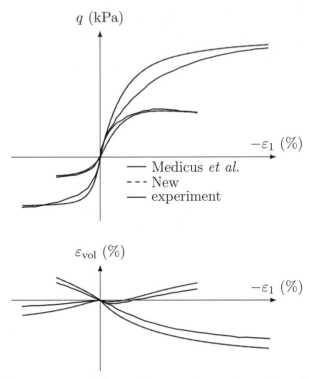

Figure 4.10: Comparison of drained triaxial test data from [91] with simulations of Barodesy in the formulation of [79] and with the new $\boldsymbol{R}$-function.

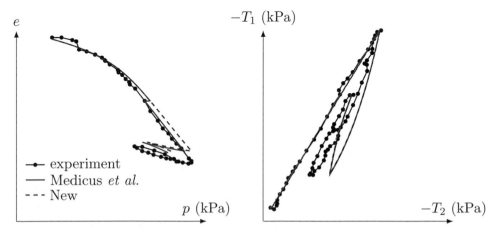

Figure 4.11: Comparison of oedometric test data from Mašín [71] with simulations of Barodesy in the formulation of [79] and with the new $\boldsymbol{R}$-function

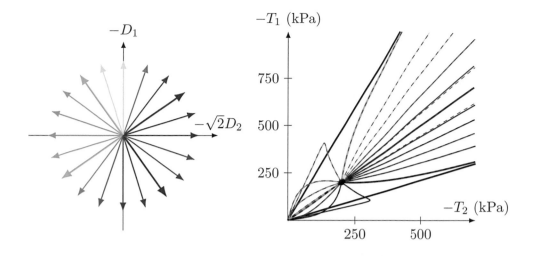

(a) Starting from an isotropic stress state

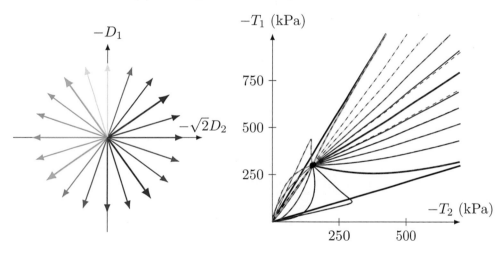

(b) Starting from an arbitrary stress state

Figure 4.12: Stress paths for proportional strain paths starting from an isotropic and an arbitrary stress state simulated with Barodesy in the formulation [79] (dashed) and with the new **R**-function (coloured)

the isotropic compression line (in this Figure $\psi_D = \pm 18°$) are slightly different for both formulations of the **R**-function.

4.2   Implementing Ohde's compression law

In this section the stiffness as defined by Ohde [90] implemented into Barodesy. Ohde [90] proposed the following equation for the stiffness for the compression of soil

$$K = K_\mathrm{r} \left(\frac{p}{p_\mathrm{r}} \right)^\xi \quad . \tag{4.20}$$

Here K_r is a reference stiffness for the stress level, p_r (which is $1\,\mathrm{kPa}$ or the atmospheric pressure) and ξ is a soil specific material parameter. Typically $\xi = 1$ for clay and between 0.5 and 1 for sand. The stiffness is also defined as

$$K = -\frac{\dot{p}}{\dot{\varepsilon}_\mathrm{vol}} \quad . \tag{4.21}$$

These equations can be combined

$$K_\mathrm{r} \left(\frac{p}{p_\mathrm{r}} \right)^\xi = K = -\frac{\dot{p}}{\dot{\varepsilon}_\mathrm{vol}} \tag{4.22}$$

which leads to a relation between volume change and stress change

$$\dot{p} = -K_\mathrm{r} \left(\frac{p}{p_\mathrm{r}} \right)^\xi \dot{\varepsilon}_\mathrm{vol} \quad . \tag{4.23}$$

In its basic form Barodesy is written as

$$\mathring{\boldsymbol{T}} = h(f\boldsymbol{R}^0 + g\boldsymbol{T}^0)\|\boldsymbol{D}\| \tag{4.24}$$

For hydrostatic compression ($T_1 = T_2 = T_3$ and without rotation $\dot{\boldsymbol{T}}{=}\mathring{\boldsymbol{T}}$ holds true) starting from a stress free point

$$\boldsymbol{R}^0 = \boldsymbol{T}^0 = \frac{-\boldsymbol{I}}{\sqrt{3}} \tag{4.25}$$

and $p = -T_1 = -T_2 = -T_3 = -T_i$ holds true, so (4.24) can be written in its diagonal components

$$\dot{T}_i = h(f+g)\frac{-1}{\sqrt{3}}\|\boldsymbol{D}\| \quad . \tag{4.26}$$

For isotropic compression deformation is defined as $D_1 = D_2 = D_3$, so the volumetric strain results in $\dot{\varepsilon}_\mathrm{vol} = \mathrm{tr}\,\boldsymbol{D} = 3D_i$ and the value of stretching $\|\boldsymbol{D}\| = -\sqrt{3}D_i$ (note, since the norm is always positive the minus here results from the negative D_i). The equations (4.23) and (4.26) can be written as

$$\dot{T}_i = K_\mathrm{r} \left(\frac{p}{p_\mathrm{r}} \right)^\xi 3D_i \quad \text{and}$$

$$\dot{T}_i = h(f+g)\frac{\sqrt{3}}{\sqrt{3}}D_i \quad .$$

The components of these two equations can be compared and it can be seen that

$$3K_\mathrm{r}\left(\frac{p}{p_\mathrm{r}}\right)^\xi = h(f+g) \quad . \tag{4.27}$$

Assuming that the stiffness term follows Ohde's law, h is

$$h = K_\mathrm{r}\left(\frac{p}{p_\mathrm{r}}\right)^\xi = K \tag{4.28}$$

it follows that the sum $f+g$ should be 3 for isotropic compression. Following Medicus and Fellin [76] for general stress states the norm of the stress $\|\boldsymbol{T}\|$ should be used. In the case of an isotropic stress state

$$\|\boldsymbol{T}\| = \sqrt{3}p \tag{4.29}$$

holds true and the stiffness term of Barodesy is then

$$h = K_\mathrm{r}\left(\frac{\|\boldsymbol{T}\|}{\sqrt{3}p_\mathrm{r}}\right)^\xi \quad . \tag{4.30}$$

In this equation the reference stiffness K_r and the reference stress p_r can be merged with the exponent ξ and the constant factor $\sqrt{3}$ as

$$c_4 = K_\mathrm{r}\left(\sqrt{3}p_\mathrm{r}\right)^{-\xi} \tag{4.31}$$

which yields

$$h = c_4\|\boldsymbol{T}\|^\xi \quad . \tag{4.32}$$

With a given equation for the stiffness an isotropic normal compression line can be derived (cf. Herle [52]). Therefore, the differential equation for the stress-strain relation has to be solved. In equation (4.23) $\dot{\varepsilon}_\mathrm{vol}$ is replaced with

$$\dot{\varepsilon}_\mathrm{vol} = \frac{\dot{e}}{1+e} \quad . \tag{4.33}$$

The differential equation is

$$\frac{\mathrm{d}p}{p^\xi} = -\frac{K_\mathrm{r}}{p_\mathrm{r}^\xi}\frac{\mathrm{d}e}{1+e} \quad , \tag{4.34}$$

with

$$-\int \frac{1}{p^\xi}\,\mathrm{d}p = \int \frac{K_\mathrm{r}}{(1+e)p_\mathrm{r}^\xi}\,\mathrm{d}e \tag{4.35}$$

and for $\xi \neq 1$ it follows

$$\frac{-p^{1-\xi}}{1-\xi} + k = \frac{K_\mathrm{r}}{p_\mathrm{r}^\xi}\ln(1+e) \quad , \tag{4.36}$$

Table 4.2: Parameter of the isotropic compression line for Hostun Sand according to Herle [51]

e_{i0}	h_s	n_H
1.09	1000 MPa	0.29

with the integration constant k. For the case of $\xi = 1$ it follows the normal compression line as proposed by Butterfield [13] which is used in Barodesy for clay [75, 76].

With (4.36) the normal compression line can be written as

$$e_i = \exp\left(-\frac{p^{1-\xi}p_r{}^\xi}{K_r(1-\xi)} + k\right) - 1 \tag{4.37}$$

Now the integration constant k has to be determined. The constant k can be excluded from the exponential function as $k_2 = \exp(k)$

$$e_i = k_2 \exp\left(-\frac{p^{1-\xi}p_r{}^\xi}{K_r(1-\xi)}\right) - 1 \tag{4.38}$$

Based on Bauer [4] for a stress free state ($T_i = 0 = p$) the void ratio should be e_{i0}. Therefore, the constant k_2 results in $1 + e_{i0}$ and the isotropic compression line is

$$e_i = (1 + e_{i0}) \exp\left(-\frac{p^{1-\xi}p_r{}^\xi}{K_r(1-\xi)}\right) - 1 \quad . \tag{4.39}$$

This expression seems quite similar to the isotropic compression line proposed by Bauer [4]

$$e_i = e_{i0} \exp\left(-\left(\frac{3p}{h_s}\right)^{n_H}\right) \tag{4.40}$$

with the material constants e_{i0}, h_s and n_H. The constants in the exponential functions can be converted from on formulation into the other

$$\xi = 1 - n_H \quad \text{and} \tag{4.41}$$

$$\frac{K_r}{p_r{}^\xi} = \frac{h_s{}^{n_H}}{3^{n_H}n_H} \quad . \tag{4.42}$$

In Fig. 4.13, the two different isotropic compression lines are compared with the parameters of Hostun sand (cf. Tab. 4.2 and 4.3). The difference between the two curves is obvious. The relative deviation with the respect to the equation of Bauer [4] is

$$\Delta e = \frac{|(1 + e_{i0})\exp(x) - 1 - e_{i0}\exp(x)|}{e_{i0}\exp(x)} = \frac{1 - \exp(x)}{e_{i0}\exp(x)} = \frac{\exp(-x) - 1}{e_{i0}} . \tag{4.43}$$

Table 4.3: Paramters of the isotropic compression line for Hostun sand calculated with (4.41) and (4.42)

e_{i0}	K_r	p_r	ξ
1.063	137.8 kPa	1 kPa	0.71

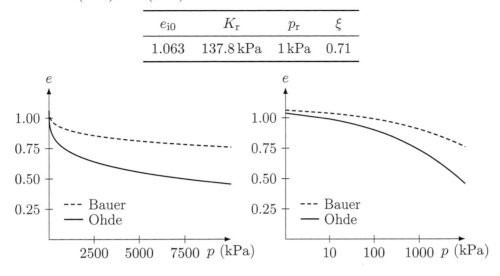

Figure 4.13: Comparison of the isotropic compression line derived from Ohde and Bauer for Hostun sand with the same material parameter from Herle [51]

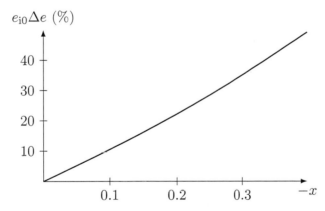

Figure 4.14: Relative error made with the same material parameters

Here x is the term in the exponent of equation (4.40) and (4.42)

$$x = -\left(\frac{3p}{h_s}\right)^{n_H} = -\frac{p^{1-\xi}p_r^{\xi}}{K_r(1-\xi)} \quad . \tag{4.44}$$

The relative deviation is shown in Fig. 4.14.

However, with a new calibration (cf. Tab. 4.4) it is possible to gain nearly the same results for the isotropic compression line, see Fig. 4.15. In Fig. 4.16,

Table 4.4: Parameter for Hostun sand so that the isotropic compression line is the same as the one according to Bauer [4]

e_{i0}	K_r	p_r	ξ
1.096	245 kPa	1 kPa	0.74

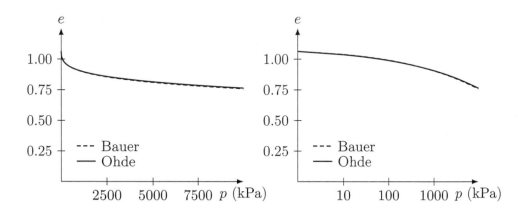

Figure 4.15: Comparison of the isotropic compression lines with different parameters

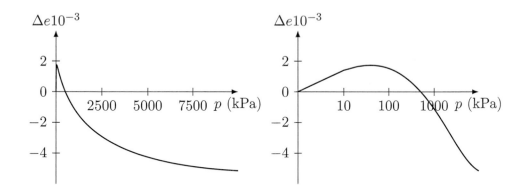

Figure 4.16: Difference between the isotropic compression lines derived from Ohde and Bauer for Hostun sand

the difference between the two isotropic compression lines is plotted. The difference is negligibly small.

A disadvantage of this function is that it is possible to get negative void ratios which makes no sense physically. However, this is just the case for large mean stresses higher than

$$p > \sqrt[1-\xi]{\frac{\ln(1+e_{i0})K_r(1-\xi)}{p_r{}^\xi}} \quad . \tag{4.45}$$

For Hostun sand with the material parameters according to Tab. 4.4 that would be for a mean stress higher than 152.5 GPa, what is far beyond the validity of the constitutive model.

Furthermore, it is assumed that the isotropic compression line is affine to the critical state line (cf. Bauer [4], Gudehus [45]), so also the critical state line is known

$$e_c = (1+e_{c0})\exp\left(-\frac{p^{1-\xi}p_r{}^\xi}{K_r(1-\xi)}\right) - 1 \tag{4.46}$$

with e_{c0} for the critical void ratio at stress free state.

The functions f, g and h build a unit; if one is changed, the others have also to be adopted. Therefore, the new h requires a redefinition of f and g. The combination of $f+g$ should include the actual void ratio e (to represent pyknotropy) and the dilatancy δ. A possible formulation for $f+g$ (following Medicus and Fellin [76]) is

$$f+g = b\delta + c_5\left[\left(\frac{1+e}{1+e_c}\right)^\zeta - 1\right] \quad , \tag{4.47}$$

here c_5 and ζ are material parameters, b is a function depending on the material and the deformation and e_c is the critical void ratio for the corresponding actual stress level.

Since the normal compression line and the critical state line are affine

$$\frac{1+e_i}{1+e_c} = \frac{1+e_{i0}}{1+e_{c0}} = c_e = \text{constant} \tag{4.48}$$

holds true and c_e can be seen as a further material parameter. This parameter determines the distance between the normal compression line and the critical state line through the void ratio (cf. Fig. 4.17a). In the clay version, this distance is for all soils twice the mean stress p (cf. Fig. 4.17b). However, it is possible to determine the parameter c_e from the material parameter λ^*

$$c_e = \frac{1+e_i}{1+e_c} = \frac{\exp\left(N - \lambda^* \ln p\right)}{\exp\left(N - \lambda^* \ln(2p)\right)} = \frac{\exp(N)\exp(-\lambda^* \ln p)}{\exp(N)\exp(-\lambda^* \ln p)\exp(-\lambda^* \ln 2)} = 2^{\lambda^*}.$$

If b is independent of stress and void ratio, the expression $f+g$ is constant for isotropic compression ($\delta = -\sqrt{3}$) starting on the normal compression line and

$$f+g = 3 = -b\sqrt{3} + c_5\left(c_e{}^\zeta - 1\right) \quad . \tag{4.49}$$

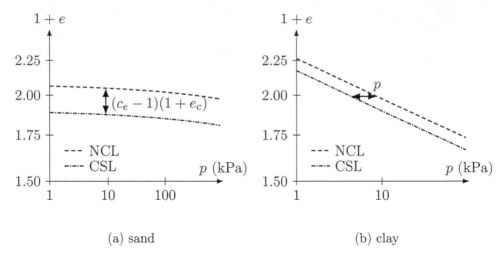

(a) sand   (b) clay

Figure 4.17: Critical state line and normal compression line for clay and sand in Barodesy

Therefore, the function b has to be for compression

$$b_{\text{comp}} = \frac{c_5 \left(c_e{}^\zeta - 1\right) - 3}{\sqrt{3}} \quad . \tag{4.50}$$

In the following it is assumed, that for isotropic extension starting from the normal compression line the stiffness is κ-times higher than the stiffness for compression. This means

$$K = \kappa K_{\text{r}} \left(\frac{p}{p_{\text{r}}}\right)^\xi = \kappa c_4 p^\xi \quad . \tag{4.51}$$

This is similar to Ohde's law and can again be compared with Barodesy. The volumetric strain rate $\dot{\varepsilon}_{\text{vol}}$ is still $3D_i$, but the value of stretching is now $\|\boldsymbol{D}\| = \sqrt{3}D_i$ since D_i is positive. The stress rates can then be written as

$$\dot{T}_i = \kappa c_4 p^\xi 3 D_i \quad \text{and} \tag{4.52}$$

$$\dot{T}_i = -h(f + g)\frac{\sqrt{3}}{\sqrt{3}}D_i \quad . \tag{4.53}$$

From isotropic compression (4.28) follows $h = c_4 p^\xi$ and so the term $f + g$ is defined as

$$-(f + g) = 3\kappa \quad . \tag{4.54}$$

With this condition, a second value for b in (4.47) is determined. For isotropic extension the dilatancy $\delta = \sqrt{3}$ and the term $f + g$ is

$$f + g = -3\kappa = b\sqrt{3} + c_5 \left(c_e{}^\zeta - 1\right) \tag{4.55}$$

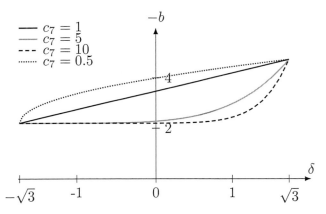

Figure 4.18: Influence of c_7 on the interpolation of b, with parameter from Tab. 4.6

and so the value of b for isotropic extension is

$$b_{\text{ext}} = \frac{c_5 \left(1 - c_e{}^\varsigma\right) - 3\kappa}{\sqrt{3}} \quad . \tag{4.56}$$

For the function $b(\delta)$ an interpolation between the values for extension and compression is used

$$b = (b_{\text{ext}} - b_{\text{comp}}) \left(\frac{\delta + \sqrt{3}}{2\sqrt{3}}\right)^{c_7} + b_{\text{comp}} \quad , \tag{4.57}$$

with the material parameter c_7. In the case of $c_7 = 1$ it is a linear interpolation between the b of extension and compression, but a higher value is found by calibration. The influence of c_7 on b can be seen in Fig. 4.18.

A further condition for the determination of $f + g$ follows from the critical state, since the stress rate is vanishing by ongoing deformation ($\|\boldsymbol{D}\| > 0$). The stress rate can just vanish either when $h = 0$ or when $f + g = 0$ and $\boldsymbol{R}^0 = \boldsymbol{T}^0$. The scalar function h is zero just in the case that the norm of the stress tensor $\|\boldsymbol{T}\|$ is zero what is not a general state. Therefore, the sum of f and g has to be zero. This requirement is already fulfilled by (4.47), since $\delta = 0$ and the void ratio is the critical void ratio ($e = e_c$)

$$f + g = b \cdot 0 + c_5 \left[\left(\frac{1 + e_c}{1 + e_c}\right)^\varsigma - 1\right] = 0 \quad . \tag{4.58}$$

This condition cannot be used to determine a material parameter. However, it shows that it was used by Medicus and Fellin [76] for constructing (4.47), such that the critical state is mapped correctly.

The term $f + g$ has to be split in his components f and g for general cases other than proportional stress paths. One possibility is

$$f = c_8 b\delta + c_6 \quad \text{and} \quad g = (1 - c_8)b\delta + c_5 \left[\left(\frac{1+e}{1+e_c} \right)^\zeta - 1 \right] - c_6 \quad . \quad (4.59)$$

According to Medicus *et al.* [79] it is assumed that for isotropic extensions the stress rate heads direct to the origin of the stress space. This means that

$$\frac{\dot{T}_i}{\dot{T}_j} = \frac{T_i}{T_j} \quad , \quad (4.60)$$

with $\boldsymbol{T} = \|\boldsymbol{T}\|\boldsymbol{T}^0$ and $\dot{\boldsymbol{T}} = \|\dot{\boldsymbol{T}}\|\mathring{\boldsymbol{T}}^0$ we obtain

$$\frac{\dot{T}_i}{\dot{T}_j} = \frac{\|\dot{\boldsymbol{T}}\|\dot{T}_i^0}{\|\dot{\boldsymbol{T}}\|\dot{T}_j^0} = \frac{\dot{T}_i^0}{\dot{T}_j^0} = \frac{T_i^0}{T_j^0} = \frac{\|\boldsymbol{T}\|T_i^0}{\|\boldsymbol{T}\|T_j^0} = \frac{T_i}{T_j} \quad . \quad (4.61)$$

Replacing $\dot{T}_i^0$ and $\dot{T}_j^0$ with (3.12) leads to

$$\frac{\dot{T}_i^0}{\dot{T}_j^0} = \frac{fR_i^0 + gT_i^0}{fR_j^0 + gT_j^0} = \frac{T_i^0}{T_j^0} \quad . \quad (4.62)$$

For isotropic extension $R_i^0 = R_j^0 = \frac{1}{\sqrt{3}}$ holds true. To full fill (4.62) in all cases, f has to be zero, what results in

$$\frac{\dot{T}_i^0}{\dot{T}_j^0} = \frac{0R_i^0 + gT_i^0}{0R_j^0 + gT_j^0} = \frac{gT_i^0}{gT_j^0} = \frac{T_i^0}{T_j^0} \quad . \quad (4.63)$$

With this condition, the constant c_8 can be determined. For isotropic extension $\delta = \sqrt{3}$ and

$$f = c_8 b_{\text{ext}} \sqrt{3} + c_6 = 0 \quad . \quad (4.64)$$

The constant c_8 is then

$$c_8 = \frac{-c_6}{\sqrt{3}b_{\text{ext}}} \quad . \quad (4.65)$$

4.3   Summary of material parameters and constants

In the here improved version of Barodesy for sand nine material parameters are used, which are

1. the critical friction angle φ_c,

2. the reference stiffness K_r,

3. the exponent ξ in the stiffness term,

4. the relation between the reference loading and unloading stiffness κ,

5. the critical void ratio at a mean stress of zero $(p = 0)$ e_{c0}

6. the ratio between the critical void ratio and the void ratio at the normal compression line at the same stress level c_e,

7. the parameter c_5 for the weighting of the pyknotropy,

8. the parameter c_6 which controls the splitting of f and g and

9. the parameter c_7, which defines the interpolation of b between isotropic compression and extension.

With these material parameters all other quantities required in Barodesy can be calculated. These quantities are

- the function b for extension and compression in f and g,

- the constants c_1, c_2 and c_3 for the $\boldsymbol{R}$-function,

- the constant c_4 in the stiffness term h,

- the constant c_8 to guaranty that the stress rate heads directly to the origin of the stress space,

- the exponent ζ in the pyknotropy term and

- the prescribed values of α in the $\boldsymbol{R}$-function (α_0, α_c and α_p).

These quantities are constants for a given soil. The equations for the constants and the constitutive model are summarised in Appendix C.1.3.

4.4   Numerical experiments

In this section results with the new formulation of Barodesy for sand should be compared with laboratory tests of Hostun sand [22], Hypoplasticity in the formulation of von Wolffersdorff [126] and the previous version of Barodesy [64] (described in sec. 3.4). The used material constants can be seen from Tab. 4.5, Tab. 4.6 and Tab. 4.7. All three models are calibrated with the set

Table 4.5: Parameter for Hostun sand for Barodesy with $p_{\text{ref}} = 1\,\text{kPa}$

φ_c °	K_r kPa	ξ	κ	e_{c0}	c_e	c_5	c_6	c_7
33.8	430	0.74	3	0.904	1.093	3	5	10

Table 4.6: Parameter for Hostun sand calibrated according to Herle [51][2] for Hypoplasticity

φ_c °	h_s MPa	n_H	e_{d0}	e_{c0}	e_{i0}	α_H	β_H
33	1000	0.29	0.63	1	1.15	0.13	2

Table 4.7: Parameter for Hostun sand calibrated for Barodesy in version of Kolymbas [64]

φ_c °	c_2	c_3	c_4 kPa	c_5	e_{c0}	$e_{\min}$
33.8	1	2.5076	1	40	0.91	0.5

of Hostun sand laboratory tests from [22] to get the best overall performance of all tests. Note, this leads to different values for Hypoplasticity and Barodesy in formulation [64] than given by Herle [51] and Kolymbas [64], respectively.

4.4.1   Isotropic compression

In Fig. 4.19, simulations of an isotropic compression test for a dense sample are shown. In the left part the stress strain curves are plotted and on the right side the void ratio e is plotted over the mean stress p. Additionally in this diagram the normal compression line and the critical state line (from Barodesy) are plotted. From the initial void ratio, it can be seen that it is a very dense sample. The new formulation of Barodesy and Hypoplasticity can reproduce the loading path very well. The previous barodetic model is too stiff. The hypoplastic model can produce also a realistic unloading, while the barodetic models is too soft in this case. For such a dense sample, the old version predicts a softer material behaviour for unloading than for loading. This causes reasonable upheaval for such a sample.

---

[2] Note, not the parameters from Herle [51] are used. The calibration was done by G. Medicus for these experiments.

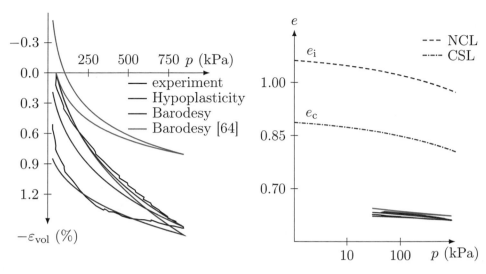

Figure 4.19: Isotropic compression test for a dense specimen

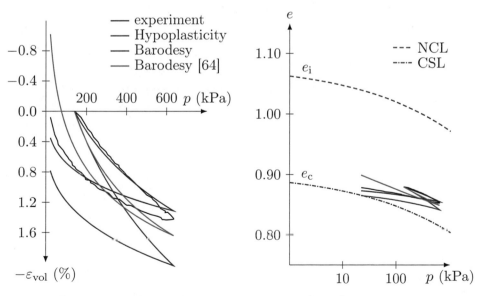

Figure 4.20: Isotropic compression test for a loose specimen

In Fig. 4.20, simulation results of a loose isotropic compression test are compared with the laboratory test. The new Barodesy shows a quite good agreement with the laboratory test, for both loading and unloading. Hypoplasticity predicts in this test too large compaction, while the unloading path shows a reasonable deformation. The previous Barodesy version is slightly too soft for loading and much too soft for unloading. This leads to a unrealistic volume increase at the end of the test.

4.4.2 Triaxial tests

The overall performance of the dense drained triaxial tests is quite good for all three constitutive models (cf. Fig. 4.21, 4.22 and 4.23). The maximum achieved deviatoric stress q is nearly the same as in the experiment and the volumetric behaviour is very similar to the experiment. However, the former version of Barodesy tends to predict too low peak friction angles and both Barodesy versions slightly overestimate the initial stiffness (especially for higher stress levels). Also the contractant behaviour at the begin of the test is not very distinct. These both defects are probably linked and could be improved with a reduction of the stiffness at the beginning. The stress curves for the unloading corresponds to the experiment. Barodesy produces for all triaxial tests a volume increase under unloading, while the experiment shows a compaction.

For the triaxial test at an initial stress level of $p_{ini} = 100\,\mathrm{kPa}$ (Fig. 4.21) the maximum deviatoric stress of the former Barodesy version is smaller than the experiment. For the other two models, the maximum deviatoric stress is nearly the same as in the experiment and the initial stiffness of all models is comparable with the stiffness of the experiment. The barodetic models show less contractancy than Hypoplasticity and the experiment at the start of the test. However, with ongoing compression the volumetric deformation of the new Barodesy version fits the volumetric behaviour better than Hypoplasticity, for which the dilatancy is too small. The former Barodesy version has even a smaller dilatancy than Hypoplasticity and clearly underpredicts the volumetric behaviour of the sample in this test.

For the other two dense triaxial tests with initial stress levels of $p_{ini} = 300\,\mathrm{kPa}$ and $p_{ini} = 600\,\mathrm{kPa}$ (Fig. 4.22 and 4.23) the results are similar. Hypoplasticity fits the stress strain curve very good. The two Barodesy versions have higher deviatoric stresses in the peak and are a little bit stiffer than the experiments and Hypoplasticity. Also in these tests Barodesy shows just a very small contractancy at the beginning, but the volumetric behaviour gets more realistic with ongoing shearing for the improved version. The previous version shows only small volumetric changes, what leads to a too small dilatancy.

Also, the simulation results for Hypoplasticity and the new Barodesy version of the loose drained triaxial test (Fig. 4.24) are similar to the laboratory test. The result gained with the previous version of Barodesy does not reach the same deviatoric stress q as the experiment or the other models. The improved Barodesy version is again stiffer at the first part of the test, while the former version and Hypoplasticity are a little bit softer than the experiment. Also in its volumetric behaviour the old version has no good prediction. The predicted compaction for this test is too large. The volumetric strain curves of the other two constitutive models follow the experimental one. Although the new Barodesy version shows in the dense test just a small contractancy for the loose sample the volumetric behaviour is very well. Also in this simulation, both Barodesy versions show a significant volume increase while unloading the sample, instead of a compaction like the experiment.

4.4.3 Oedometric tests

The results of the oedometric tests are shown in Fig. 4.25 and 4.26. Hypoplasticity and the new barodetic model show nearly the same stress-strain curves for loading, which are initial softer than the experiment. Nevertheless, at the end of the loading they show nearly the same deformation as the experiment. The old Barodesy version is even softer than the other two models and predicts a too large deformation. Hypoplasticity shows a correct unloading stiffness (Fig. 4.25). The barodetic models show a too low stiffness for unloading. For the new version even a heaving occurs. In Fig. 4.25b, also the evolutions of the stress components are shown. For the loading, it can be seen that the increase of the lateral stress is much higher for the former Barodesy version and follows not a K_0-path. For the new Barodesy version, there is an increase of the radial stress $T_2 = T_3$ at the start of the unloading. This behaviour is mechanical not realistic for soil. The stress curves of Hypoplasticity and the former Barodesy version seem to give a realistic stress response for an oedometric unloading.

The simulations of the loose oedometric experiment show different behaviours for all three models, Fig. 4.26. Hypoplasticity produces again a slightly too large compaction, while the new developed Barodesy version is a little bit too stiff and the former Barodesy version is much too soft. With the previous version of Barodesy, it is only possible to fit a dense or a loose sample quite well. For this test Hypoplasticity and the improved Barodesy version show realistic unloading stiffness in contrast to the former version of Barodesy.

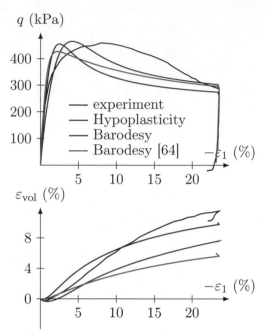

Figure 4.21: Drained triaxial test for a dense specimen with a consolidation pressure of 100 kPa data from Desrues *et al.* [22]

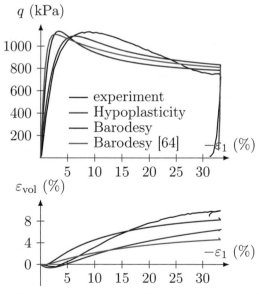

Figure 4.22: Drained triaxial test for a dense specimen with a consolidation pressure of 300 kPa data from Desrues *et al.* [22]

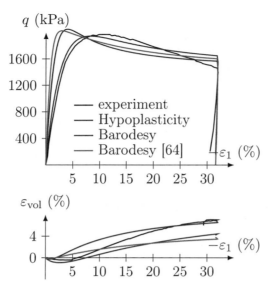

Figure 4.23: Drained triaxial test for a dense specimen with a consolidation pressure of 600 kPa data from Desrues *et al.* [22]

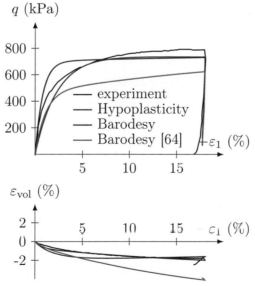

Figure 4.24: Drained triaxial test for a loose specimen with a consolidation pressure of 300 kPa data from Desrues *et al.* [22]

However, the new Barodesy version shows again the unrealistic initial increase of the radial stress $T_2 = T_3$ for unloading.

4.4.4   Response envelopes and proportional strain paths

Additional to the simulation of the laboratory tests, response envelops and proportional strain paths are plotted to proof the quality of the improved Barodesy. Response envelops were introduced by Gudehus [44] to describe and compare different constitutive models. For a given admissible stress state in the Rendulic plane the stress rates for all stretchings with a normed stretching of one ($\|\boldsymbol{D}\| = 1$) are calculated. These stress rates are scaled and plotted into the Rendulic plane (cf. Fig. 4.27 and 4.28) to see the direction and the relative size of a stress increment.

To justify the quality of the response envelops also response envelops for Hypoplasticity are shown. Due to the formulation of Hypoplasticity the response envelops are ellipses. In Fig. 4.27, the void ratio is selected to correspond to the void ratio at the normal compression line for this stress level. Due to different approaches for the isotropic compression line the void ratios differ in this plot.

It can be seen that the response envelops of Barodesy are larger than the one of Hypoplasticity, i.e. Barodesy is stiffer than Hypoplasticity. This result is in good accordance with the results of the loose isotropic compression test, where Hypoplasticity was slightly too soft. The shape of the response envelops for Barodesy is acceptable. The loading and unloading stiffness for the isotropic test is different from each other and the stress rate for isotropic extension points in the direction of the origin.

In Fig. 4.28, the response envelops for a dense sample is shown. The void ratio is the same for both samples in this case and two stress points on the Matsuoka-Nakai failure surface are chosen. For these initial conditions, the constitutive models show a similar stiffness. It is possible to exceed the cone defined by the Matsuoka-Nakai failure criteria with Barodesy. Note, this is just the case for dense samples.

In Fig. 4.29 and 4.30 starting from a given stress state proportional strain paths are applied for Barodesy and Hypoplasticity. From these figures it can be seen whether the stress paths leave the permissible stress region or not. For soils, the negative stress octant is a permissible region. Additionally the proportional stress paths for isochoric deformations ($\delta = 0$) starting from a stress free point are plotted in black and coincide with the Matsuoka-Nakai failure surface. For loose samples, all stress paths have to remain between

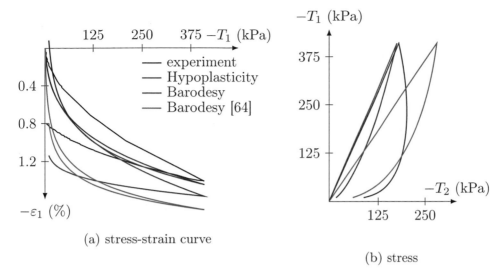

(a) stress-strain curve

(b) stress

Figure 4.25: Oedometric test for a dense Hostun sand specimen data from Desrues *et al.* [22]

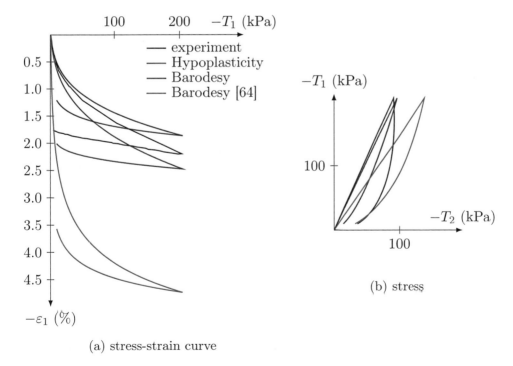

(a) stress-strain curve

(b) stress

Figure 4.26: Oedometric test for a loose Hostun sand specimen data from Desrues *et al.* [22]

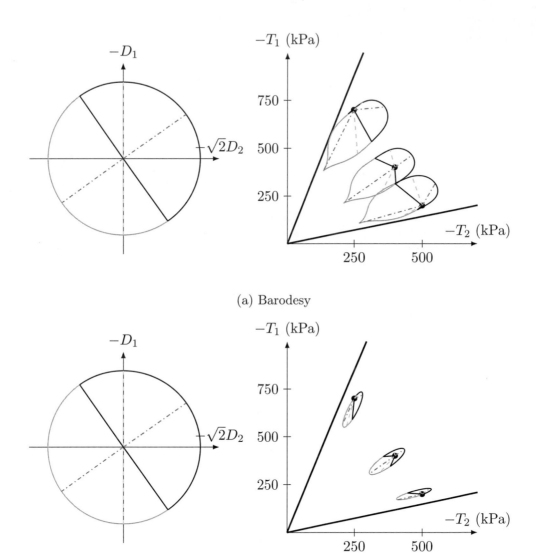

(a) Barodesy

(b) Hypoplasticity

Figure 4.27: Response envelopes for a loose sample with $e = e_i$

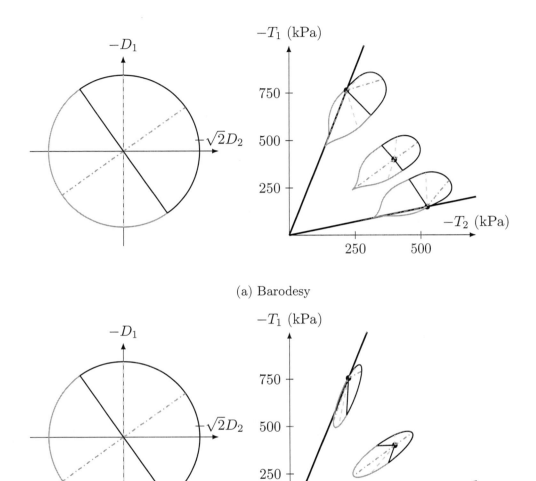

(a) Barodesy

(b) Hypoplasticity

Figure 4.28: Response envelopes for a dense sample with $e = 0.75$

these two lines. Stress paths of dense samples with dilatant strain paths are able to exceed this range, which results in mobilised friction angles higher than the critical friction angle.

The stress paths starting from an isotropic stress state are shown in Fig. 4.29. The sample has a void ratio at start of $e = 0.75$, i.e. a dense sample. Barodesy shows good results. It is possible to reach mobilised friction angles higher than the critical one for dilatant strain paths and the isochoric deformation path approaches the one from the stress free state.

One shortcoming of Hypoplasticity can be seen in this plot. For an isochoric strain path the stress state exceeds the Matsuoka-Nakai surface. This can also be seen in experiments, but the stress state returns back on this failure surface immediately. The stress path of Hypoplasticity returns back to the surface too. The difference is that this happens at an unrealistic high mean stress, because the surface is reached in the critical state. For the here shown case with $e = 0.75$ the surface is reached at a mean stress level of $p = 899\,\text{MPa}$.

In Fig. 4.30, the results of loading with proportional strain paths starting at a more general stress state are shown. Again, the initial void ratio is $e = 0.75$. The outcome is comparable with the previous one. Barodesy shows mobilised friction angles higher than the critical ones for dilatant strain paths. However, Barodesy predicts no phase transition, contrary to Hypoplasticity.

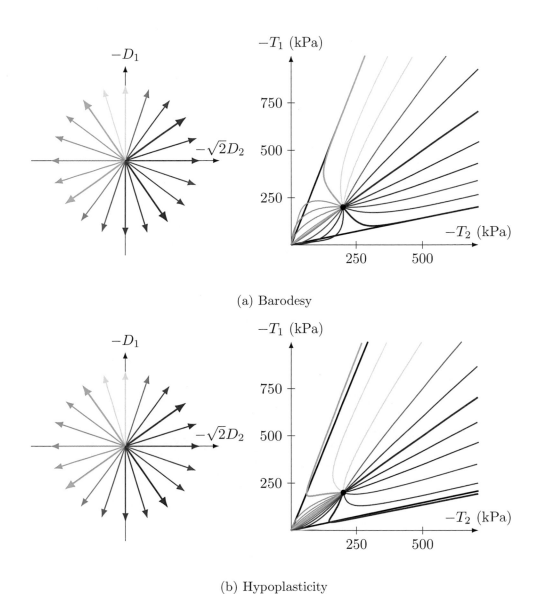

(a) Barodesy

(b) Hypoplasticity

Figure 4.29: Stress paths for different proportional strain paths starting from an isotropic stress state with a void ratio $e = 0.75$

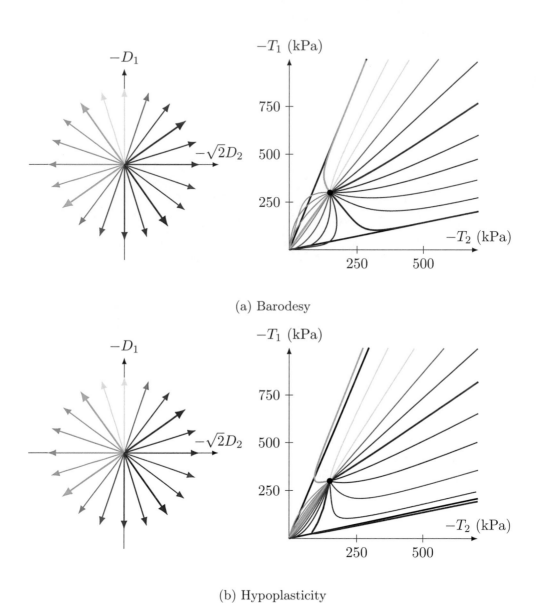

(a) Barodesy

(b) Hypoplasticity

Figure 4.30: Stress paths for different proportional strain paths starting from a non-isotropic stress state and a void ratio of $e = 0.75$

Chapter 5

Deformations induced by principal stress rotation

5.1 Introduction

Soil is often exposed to a rotation of the principal stress axes due to engineering work. Even in the case of a simple foundation with only vertical load a rotation of the principal axes will be imposed (except under the centre). The foundations of offshore platforms are exposed to a rotation of the principal axes due to wave action, as Ishihara and Towhata [55] showed. Another example is deep vibro-compaction. The vibroflot emits a combination of compression and shear waves [28], which also induce principal stress rotations in the soil. As compaction is the main issue here, a material model used in a computation must predict volume changes due to principal stress rotations.

In conventional laboratory tests (e.g. triaxial test, oedometer test), such a rotation of principal axes is not feasible. Nowadays several apparatus have been constructed to perform tests on soil, where a rotation of the principal axes is possible (e.g. the hollow cylinder apparatus [55], the directional shear cell [3], or the $1\gamma2\varepsilon$ apparatus [58]). In experiments with these apparatus a volumetric change of the sample occurs although the stress invariants remain constant.

In the following, the behaviours of several constitutive relations (elastoplastic, hypoplastic and barodetic models) are compared with results from a $1\gamma2\varepsilon$ apparatus and a hollow cylinder apparatus.

5.2 Models

The following constitutive relations are investigated: Sanisand [110], Hardening Soil with small-strain stiffness [7], Hypoplasticity [61, 63] in the formulation of von Wolffersdorff [126] with and without the extension of Niemunis and

Herle [88] for intergranular strain ("Hypo i.s.") and Barodesy in the version
of Kolymbas [64] and the improved version as described in chapter 4.

5.2.1 Hardening Soil with small-strain stiffness

The Hardening Soil with small-strain stiffness model (Benz [7]) is an extension
of Hardening Soil developed by Schanz [100] and Schanz et al. [101]. It is an
elastoplastic constitutive relation with shear strain and compression hardening. This model does not use the void ratio e as a state variable and hence
does not include the concept of the critical state soil mechanics.

The model produces a nonlinear stiffness for an initial loading, due to the
hardening of the yield surface. For a calculation starting from an isotropic
state the yield surface coincides with the isotropic axis. With ongoing shearing
the yield surface increases until the yield surface reaches the strength criteria,
i.e. the Mohr-Coulomb failure criterion. The yield surface for shearing under
triaxial conditions can be described by [7]

$$f_{\mathrm{y}}^{\mathrm{s}} = \frac{q_{\mathrm{a}}}{E_{50}} \frac{T_3 - T_1}{q_{\mathrm{a}} - (T_3 - T_1)} - \frac{2(T_3 - T_1)}{E_{\mathrm{ur}}} - \gamma^{\mathrm{ps}} \quad , \tag{5.1}$$

where E_{50} and E_{ur} are material stiffness moduli for triaxial compression and
unloading/reloading, respectively (which depend on the minor principal stress
T_3), q_{a} is the asymptotic deviatoric stress for a triaxial test as introduced in
Duncan and Chang [25] and γ^{ps} is the accumulated plastic shear strain. In
the case of unloading, as for the reloading until the yield surface is reached
again, the model reacts elastic.

To describe isotropic loading realistically, the model uses a cap for the yield
surface, which is described by

$$f_{\mathrm{y}}^{\mathrm{c}} = \frac{\tilde{q}^2}{\tilde{\alpha}^2} - p^2 - p_{\mathrm{p}}^2 \quad . \tag{5.2}$$

In this equation $\tilde{q}$ is a stress measurement to adjust the cap to the yield
surface for shearing, $\tilde{\alpha}$ is in material parameter, and p_{p} is the preconsolidation
pressure, which is the hardening parameter.

The small-strain stiffness is introduced with a small-strain overlay into the
model. For small strain cycles, the shear stiffness depends only on the strain
history, whereas for larger strains it is stress dependent.

Additionally, in the small-strain model the definition of the mobilized dilatancy angle is changed, so that also negative mobilised friction angles are

possible. With this change to the original model the contractive behaviour is improved [7].

For more details of the model, please see Benz [7].

5.2.2 Sanisand

Sanisand (Simple ANIsotropic SAND plasticity model) is the name for a family of constitutive models, which were developed over the past decades (Dafalias and Manzari [19], Dafalias *et al.* [20], Manzari and Dafalias [70]). These elasto-plastic models incorporate the framework of critical state soil mechanics and the bounding surface plasticity.

The model used in this chapter is from Taiebat and Dafalias [110]. It consists of a narrow cone that forms the yield surface and three other surfaces in the stress space. They define the stress states in the critical state, a limit surface for admissible stress states, as well as a dilatancy surface. Furthermore, it incorporates a critical state line and a limiting compression line.

To decide whether the sand is loose or dense in the sense of critical state soil mechanics the critical state line as proposed by Li and Wang [68]

$$
e_{\mathrm{c}} = e_{c0} - \lambda_e \left(\frac{p}{p_{\mathrm{at}}} \right)^{\xi_e} \quad ,
\tag{5.3}
$$

where e_{c0}, λ_e and ξ are material parameters and p_{at} is the atmospheric pressure, is used. With the critical void ratio for the current stress state the state parameter $\psi_e = e - e_c$, originally defined by Been and Jefferies [6], can be determined, Fig. 5.1. A value $\psi_e > 0$ describes loose sand, where $\psi_e < 0$ describes dense sand and in a critical state $\psi_e = 0$ applies. A limiting compression curve similar to the one proposed by Pestana and Whittle [92] is incorporated via the hardening for very high stress levels.

The strain is decomposed into a volumetric part, $\varepsilon_{\mathrm{vol}} = \mathrm{tr}\,\boldsymbol{\varepsilon}$, and a deviatoric part

$$
\boldsymbol{\varepsilon}^* = \boldsymbol{\varepsilon} - \varepsilon_{\mathrm{vol}} \frac{\boldsymbol{I}}{3} \quad .
\tag{5.4}
$$

In the same way the stretching is separated into a volumetric part D_{vol} and a deviatoric part $\boldsymbol{D}^*$ and the stress is decomposed into the mean stress p and the deviatoric stress tensor

$$
\boldsymbol{T}^* = \boldsymbol{T} - p\boldsymbol{I} \quad .
$$

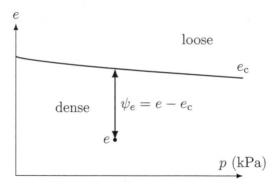

Figure 5.1: Critical state line and state parameter ψ_e

The elastic part of the model is defined by the following relation between stress rate and stretching

$$D^e_{\text{vol}} = \frac{\mathring{p}}{K} \quad \text{and} \quad \boldsymbol{D}^{*e} = \frac{\mathring{\boldsymbol{T}}^*}{3G} \quad , \tag{5.5}$$

here the superscript e denotes the elastic part of the stretching. The model uses the hypoelastic bulk modulus

$$K = K_r p_{\text{at}} \frac{1+e}{e} \left(\frac{p}{p_{\text{at}}} \right)^{2/3} \quad , \tag{5.6}$$

and the hypo-elastic shear modulus

$$G = G_0 p_{\text{at}} \frac{(2.97 - e)^2}{1+e} \sqrt{\frac{p}{p_{\text{at}}}} \tag{5.7}$$

with the model parameters K_r and G_0, and the atmospheric pressure p_{at}. The pure elastic range of the model is quite small, since the yield surface is a quite narrow range (Fig. 5.4) with an apex at the origin and a cap, which can move in the stress space. This yield surface is described by

$$f_y = \frac{3}{2} \left(\boldsymbol{T}^* - p\boldsymbol{\alpha} \right) : \left(\boldsymbol{T}^* - p\boldsymbol{\alpha} \right) - m_S^2 p^2 \left[1 - \left(\frac{p}{p_0} \right)^{n_S} \right] \tag{5.8}$$

where the hardening parameters $\boldsymbol{\alpha}$ and p_0 define the direction of the axis of the yield surface and the top of the cap of the yield surface, respectively, and m_S is the tangent of the half-opening angle of the yield surface. The shape of the yield surface in the p-q-plane can be seen in Fig. 5.2, in the deviatoric plane it is circle in the range of $0 < p < p_0$, cf. Fig. 5.4.

In the case of yielding, the yield surface f_y can move in the stress space and the axis $\boldsymbol{\alpha}$ is limited by the bounding surface f^b. The bounding surface has

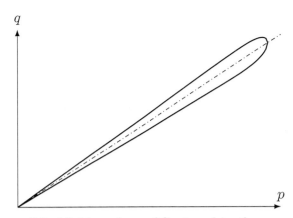

Figure 5.2: Yield surface of Sanisand in the p-q-plane

the same shape as the critical surface f^c and the dilatancy surface f^d. The shape and size of the critical surface f^c is constant, whereas the size of the other two surfaces changes depending on ψ_e.

Taiebat and Dafalias [110] use a shape for f^c, which is based on the proposition by Argyris et al. [2]. In the deviatoric plane the distance from the origin is defined with the following equation

$$f^c(\theta) = \frac{2c_\alpha}{(1 + c_\alpha) - (1 - c_\alpha)\cos 3\theta} p\alpha_c^c \quad . \tag{5.9}$$

In this equation θ, c_α and α_c^c denotes the Lode angel, the relation between the stress ratio in critical state for triaxial compression and extension and the stress ratio in critical state for triaxial compression, respectively. Note, this surface is not convex for all relevant critical stress ratios, however this raises no problems since the surface is not used as plastic potential. The shape of the critical surface f^c in the deviatoric plane can be seen in Fig. 5.3. In this and the following figures stresses normalised by the mean stress p will be used $(\boldsymbol{r} = \boldsymbol{T}/p)$. For a dense soil ($\psi_e < 0$) the bounding surface f^b is larger than the critical surface and the dilatancy surface f^d is smaller (Fig. 5.3a), in the case of loose soil ($\psi_e > 0$) the dilatancy surface f^d is larger and the bounding surface f^b is smaller than the critical surface (Fig. 5.3b). In the critical state, all three surfaces coincide (5.4f).

The dilatancy surface is used to determine, whether volumetric plastic strain leads to compaction or loosening. For a dense sample in a triaxial test, as shown in Fig. 5.4g, the volumetric plastic stretching (like the elastic stretching) leads to a decrease of void ratio until the back-stress ratio $\boldsymbol{\alpha}$ crosses the dilatancy surface. During this process, the bounding surface grows, since ψ_e decrease, and the dilatancy surface shrinks, Fig. 5.4a. After the dilatancy surface is crossed the volumetric plastic stretching causes the void ratio increase

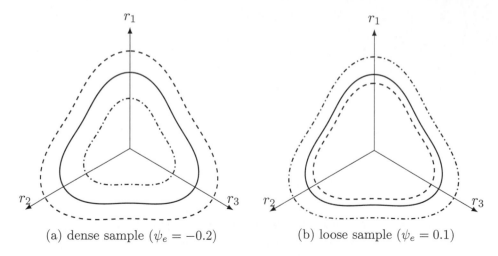

(a) dense sample ($\psi_e = -0.2$)   (b) loose sample ($\psi_e = 0.1$)

Figure 5.3: Illustration of the surfaces in Sanisand in the deviatoric plane for different void ratios

what causes ψ_e to decrease. This results in a shrinkage of the bounding surface and a growth of the dilatancy surface, Fig. 5.4b. For a dense sample, which is able to reach a peak friction angle φ_p higher than the critical friction angle φ_c, it is possible to reach a stress state beyond the critical surface, Fig. 5.4c. However, with ongoing shearing the back stress ratio $\boldsymbol{\alpha}$ will reach the bounding surface f^b, Fig. 5.4d, and with ongoing shearing the bounding surface push the back-stress ratio into the direction of the critical surface, Fig. 5.4e. After a very long shearing, when the sample is in the critical state, the bounding and the dilatancy surfaces coincide with the critical surface, since $\psi_e = 0$, Fig. 5.4f.

In the case of a loose sample, the bounding surface is smaller than the critical surface, i.e. the range of possible stress states is smaller than in the critical state. However, with ongoing shearing the soil compacts and the void ratio approaches the critical void ratio. The bounding surface increases during this process, until it has the same size as the critical surface.

All equations, which are required for implementation of the model can be found in Appendix C.3. For further details on the constitutive model (hardening rules and the plastic potential), please see Taiebat and Dafalias [110].

5.2.3 Hypoplasticity

Hypoplasticity is the name of a family of constitutive models. The first version of a hypoplastic model has been introduced by Kolymbas [60]. Since then,

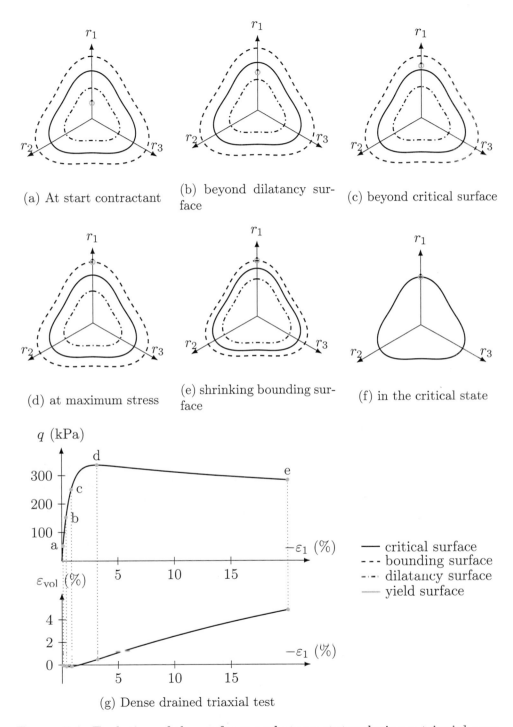

Figure 5.4: Evolution of the surfaces and stress states during a triaxial compression test

many different versions and improvements have been published (cf. Kolymbas and Medicus [66]).

An important difference between elastoplastic and hypoplastic constitutive models is that the hypoplastic models use one equation to calculate the objective stress rate $\overset{\circ}{\boldsymbol{T}}$ and do not distinguish between elastic and plastic deformations. Hence, Hypoplasticity needs no yield surface and no plastic potential. In general, Hypoplasticity can be written as

$$\overset{\circ}{\boldsymbol{T}} = h(\boldsymbol{T}, \boldsymbol{D}, \dots) = \mathcal{L} : \boldsymbol{D} + \boldsymbol{N} \|\boldsymbol{D}\| \quad . \tag{5.10}$$

Here $\mathcal{L}$ denotes a forth-order tensor for the linear term and $\boldsymbol{N}$ a second order tensor for the nonlinear term of the hypoplastic function h. Due to a linear and a nonlinear term, the stiffness for loading and unloading is different. The models are homogeneous of degree n in stress $\boldsymbol{T}$, i.e. the stiffness is stress-dependant, and positive homogeneous of degree one in the rate of deformation $\boldsymbol{D}$, i.e. these versions are rate independence. Furthermore in all recent models the void ratio e is an additional variable in the equations, with this variable it is possible to incorporate the concepts from critical state soil mechanics. In this work the Hypoplasticity version for sand of von Wolffersdorff [126] is used.

5.2.4 Hypoplasticity with intergranular strain

One of the principal shortcomings of Hypoplasticity in the version of von Wolffersdorff [126] (and other versions) is the large accumulation of deformation for small stress cycles, which is called ratcheting. Niemunis and Herle proposed a small elastic range for the case of stress or strain changes to improve this behaviour.

They split the strain into an intergranular part, which is described as an intergranular interface layer, and a grain rearranging part. The intergranular part is modelled by the new state variable $\boldsymbol{\delta}$. The norm of the intergranular strain is limited to the material parameter R. The evolution of $\boldsymbol{\delta}$ is given by

$$\dot{\boldsymbol{\delta}} = c_\eta \left(\boldsymbol{D} - \frac{\boldsymbol{\delta} \|\boldsymbol{D}\|}{R} \right) \quad , \tag{5.11}$$

where the material constant c_η controls the approach rate to the maximum R.

After a constant deformation, when the norm of intergranular strain $\boldsymbol{\delta}$ has reached its maximum R, three different cases can be distinguished. For an ongoing deformation, the intergranular strain has then no influence. The second case is a reversal of the loading direction. Here the nonlinear term of Hypoplasticity is switched off and the stiffness of the linear term is increased by the new material constant m_r. In the last case the new deformation direction is orthogonal to the previous deformation direction. Here, too, the nonlinear term is switched off and the stiffness of the linear term is increased. However, in this case not with the factor m_r but with m_t, which is smaller than m_r. A change of the deformation direction which is between these cases is handled by an interpolation.

For the case that the intergranular strain δ is zero, the response of the material is elastic in all directions with the same stiffness as for deformation reversal. In the cases $0 < \|\delta\| < R$ a nonlinear interpolation is made. The equations of the intergranular strain concepts can be found in App. C.2.2.

For more information about the model, numerical aspects and the calibration of the additional constants, see [88].

5.2.5   Calibration

To compare the different models in a first step various laboratory tests are simulated: a drained triaxial test, a simple shear test with constant volume and a simple shear test with constant normal stress are simulated with Toyura sand [125]. The material parameters can be seen in Tab. 5.1, Tab. 5.2, Tab. 5.3, Tab. 5.4 and Tab. 5.5. For Sanisand and Hypoplasticity, parameters from literature ([110] and [51]) are used. Hardening Soil and Barodesy are calibrated with results of a triaxial test from [125].

The consolidation pressure of the triaxial test is 100 kPa and the initial void ratio e_0 after the consolidation is 0.831. The results of the numerical computation can be seen in Figure 5.5. Taking into account the experimental scatter [107] all models perform quite well.

In the simple shear test with constant volume the sample is loaded with a vertical load of 100 kPa, the lateral pressure is set to 50 kPa (which corresponds to a K_0-state). The initial void ratio is the same as in the triaxial test ($e_0 = 0.831$). The stress evolution is shown in Fig. 5.6. The differences are more pronounced than for the drained triaxial test.

For the simple shear with a constant vertical stress, the initial conditions are the same as in the simple shear test with constant volume. In Fig. 5.7, the results of this test are shown. The differences of the stress evolution are in the same order as in the drained triaxial test. Note, that except for Hardening Soil Small the volumetric changes are very similar in this test.

Table 5.1: Material parameters of Sanisand for Toyura Sand [110]

G_0 kPa	K_0 kPa	α_c^c	c_α	e_{c0}	λ_e	ξ_e	n^d	A_d	n^b	h_0	c_h	p_r MPa	p_c	θ_L	X
125	150	1.2	0.712	0.934	0.019	0.7	2.1	0.4	1.25	36.96	0.987	5.5	0.37	0.18	0.8

Table 5.2: Material parameters of Hardening Soil Small (calibrated for Toyura Sand from [125])

E_{50}^{ref} MPa	E_{oed}^{ref} MPa	E_{ur}^{ref} MPa	m_H	c kPa	φ °	ψ °	ν_{ur}	K_0^{nc}	R_f	σ_{Tension} kPa	E_0^{ref} MPa	$\gamma_{0.7}$
30	30	120	0.6	0	33	4	0.2	0.46	0.9	0	300	0.0002

Table 5.3: Material parameters of Hypoplasticity in the version of von Wolffersdorff [51] and the parameters of intergranular strain

φ_c °	h_s GPa	n_H	e_{d0}	e_{c0}	e_{i0}	α_H	β_H
30	2.6	0.27	0.61	0.98	1.10	0.18	1.00

Table 5.4: Material parameters of Barodesy Kolymbas [64] (preliminary calibration for Toyura Sand from [125])

φ_c °	c_2	c_3	c_4	c_5	e_{c0}	$e_{\min}$
30	1.3	−1.5	2	40	0.93	0.55

Table 5.5: Material parameters of Barodesy in its here improved version (preliminary calibration for Toyura Sand from [125])

φ_c °	K_r kPa	ξ	κ	e_{c0}	c_e	c_5	c_6	c_7
30	1000	0.6	4	0.95	1.059	2	6	8

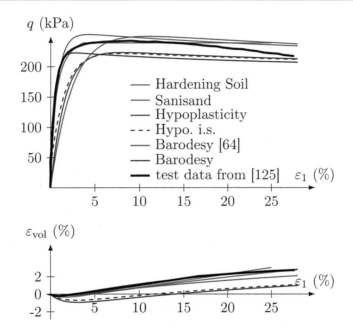

Figure 5.5: Stress-strain and volumetric strain curves for different constitutive models compared with a triaxial test in [125]

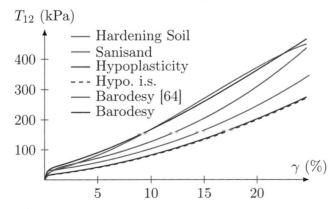

Figure 5.6: Stress-strain curves for a simple shear test with constant volume obtained with different constitutive models

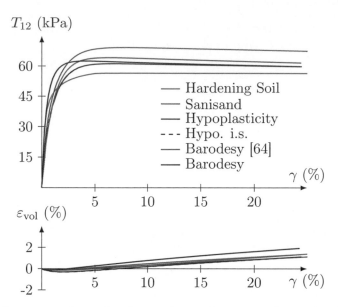

Figure 5.7: Stress-strain and volumetric strain curves for a simple shear test with constant normal stress obtained with different constitutive models

5.3   Kinematic analysis of the deformations

5.3.1   Plane strain

In a first step the deformation in a two dimensional case is investigated, before the general three-dimensional case is developed. This case coincides with plane strain conditions in three dimensions, where deformations in the third direction are constrained, i.e. $x_3(t) = X_3$. This allows to reduce the matrices to two dimensions. Fig. 5.8 shows a possible definition of the deformation. Here the function $a_1(t)$ describes a shearing in the direction x_1, and the functions $a_2(t)$ and $a_3(t)$ describe elongations in the x_2 and x_1 direction. Note, a possible shearing in the direction x_2 is not considered with respect to the deformation in the $1\gamma2\varepsilon$ apparatus. The current deformed configuration $\boldsymbol{x}(t)$ can be described by the functions

$$x_1(t) = X_1 + X_1 a_3(t) + X_2 a_1(t) \quad \text{and} \tag{5.12}$$
$$x_2(t) = X_2 + X_2 a_2(t) \quad . \tag{5.13}$$

The undeformed reference configuration (the state at $t = 0$) is described here with $\boldsymbol{X}$.

The deformation gradient $\boldsymbol{F}$ is

$$\boldsymbol{F} = \frac{\partial \boldsymbol{x}}{\partial \boldsymbol{X}} = \begin{bmatrix} 1 + a_3 & a_1 \\ 0 & 1 + a_2 \end{bmatrix} \quad . \tag{5.14}$$

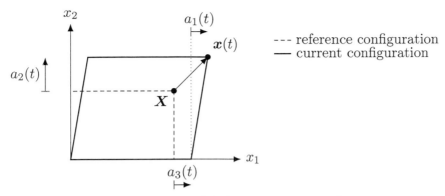

Figure 5.8: Kinematic of deformation in two dimensions

With the inverse

$$F^{-1} = \begin{bmatrix} \frac{1}{1+a_3} & \frac{-a_1}{(1+a_3)(1+a_2)} \\ 0 & \frac{1}{1+a_2} \end{bmatrix} \quad , \tag{5.15}$$

the velocity gradient L can be determined

$$L = \frac{\partial \dot{x}}{\partial x} = \dot{F} F^{-1} = \begin{bmatrix} \frac{\dot{a}_3}{1+a_3} & \frac{\dot{a}_1(1+a_3)-\dot{a}_3 a_1}{(1+a_3)(1+a_2)} \\ 0 & \frac{\dot{a}_2}{1+a_2} \end{bmatrix} \quad . \tag{5.16}$$

With the velocity gradient, the rate of deformation D

$$D = \frac{1}{2}\left(L + L^{\mathsf{T}}\right) = \begin{bmatrix} \frac{\dot{a}_3}{1+a_3} & \frac{\dot{a}_1(1+a_3)-\dot{a}_3 a_1}{2(1+a_3)(1+a_2)} \\ \frac{\dot{a}_1(1+a_3)-\dot{a}_3 a_1}{2(1+a_3)(1+a_2)} & \frac{\dot{a}_2}{1+a_2} \end{bmatrix} \tag{5.17}$$

and the spin tensor W

$$W = \frac{1}{2}\left(L - L^{\mathsf{T}}\right) = \begin{bmatrix} 0 & \frac{\dot{a}_1(1+a_3)-\dot{a}_3 a_1}{2(1+a_3)(1+a_2)} \\ \frac{\dot{a}_3 a_1-\dot{a}_1(1+a_3)}{2(1+a_3)(1+a_2)} & 0 \end{bmatrix} \tag{5.18}$$

can be determined.

5.3.2 Analysis in a plane stress state (three dimensions)

To describe a rotation of the principal stresses all three dimension have to be taken into account to keep the out of plane stress T_{33} constant.

As in the two dimensional case $a_1(t)$ describes the shearing in the x_1-x_2-plane in direction of x_1 and the elongations in x_1 and x_2 direction are described by $a_3(t)$ and $a_2(t)$, respectively. For a constant normal stress T_{33} in x_3 direction, the possibility of deformation in this direction has to be introduced. This is

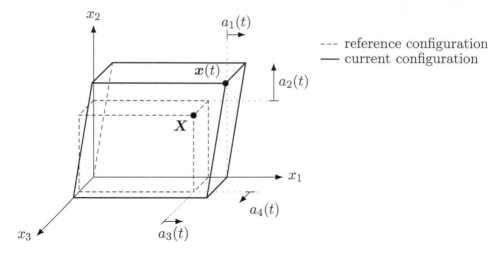

Figure 5.9: Kinematic of deformation in three dimensions

done with the function $a_4(t)$. Further functions (e.g. for shearing in an x_3-plane) are not necessary to describe the deformation during stress rotation with a constant principal stresses. In Fig. 5.9, a deformed body with the definition of the functions is shown.

The deformation can be described by the functions

$$
\begin{aligned}
x_1(t) &= X_1 + X_1 a_3(t) + X_2 a_1(t) \quad , \\
x_2(t) &= X_2 + X_2 a_2(t) \quad \text{and} \\
x_3(t) &= X_3 + X_3 a_4(t) \quad .
\end{aligned}
\tag{5.19}
$$

The stretching tensor $\boldsymbol{D}$ of this deformation is

$$
\boldsymbol{D} =
\begin{bmatrix}
\dfrac{\dot{a}_3}{1+a_3} & \dfrac{\dot{a}_1(1+a_3)-\dot{a}_3 a_1}{2(1+a_3)(1+a_2)} & 0 \\[3mm]
\dfrac{\dot{a}_1(1+a_3)-\dot{a}_3 a_1}{2(1+a_3)(1+a_2)} & \dfrac{\dot{a}_2}{1+a_2} & 0 \\[3mm]
0 & 0 & \dfrac{\dot{a}_4}{1+a_4}
\end{bmatrix}
\tag{5.20}
$$

and the spin tensor $\boldsymbol{W}$ is

$$
\boldsymbol{W} =
\begin{bmatrix}
0 & \dfrac{\dot{a}_1(1+a_3)-\dot{a}_3 a_1}{2(1+a_3)(1+a_2)} & 0 \\[3mm]
\dfrac{\dot{a}_3 a_1-\dot{a}_1(1+a_3)}{2(1+a_3)(1+a_2)} & 0 & 0 \\[3mm]
0 & 0 & 0
\end{bmatrix} \quad ,
\tag{5.21}
$$

which is quite similar to the result in two dimensions.

5.4 Calculation of the stress and strain rates

The objective stress rate $\overset{\circ}{\boldsymbol{T}}$, which is delivered by constitutive relations, is expressed by means of the Jaumann-Zaremba rate [120]

$$\overset{\circ}{\boldsymbol{T}} = \dot{\boldsymbol{T}} - \boldsymbol{W}\boldsymbol{T} + \boldsymbol{T}\boldsymbol{W} \quad , \tag{5.22}$$

where $\dot{\boldsymbol{T}}$ is the time derivative of the stress and $\boldsymbol{T}$ is the actual Cauchy-stress.

Note, there are also other stress rates, e.g. the Green-Naghdi stress rate [43]

$$\overset{\circ}{\boldsymbol{T}} = \dot{\boldsymbol{T}} - \boldsymbol{\Omega}\boldsymbol{T} + \boldsymbol{T}\boldsymbol{\Omega} \quad , \tag{5.23}$$

which uses the angular velocity tensor $\boldsymbol{\Omega}$ instead of the spin tensor $\boldsymbol{W}$. The angular velocity tensor is calculated with the rotation tensor $\boldsymbol{Q}$ and the rotation rate tensor $\dot{\boldsymbol{Q}}$ as $\boldsymbol{\Omega} = \dot{\boldsymbol{Q}}\boldsymbol{Q}^\mathsf{T}$. This stress rate causes no stress oscillations for large deformations like the Jaumann-Zaremba rate as has been shown by Bauer [5] and Dienes [23].

In the following the stress rate according to Jaumann and Zaremba will be used, since the calculation of the spin tenor $\boldsymbol{W}$ is much easier than the angular velocity tensor $\boldsymbol{\Omega}$ (for the calculation of $\boldsymbol{Q}$ see the end of this section) and the shear deformations are smaller than the limit defined by Dienes [23] for which significantly different stresses occur (the limit for a simple shear calculation is defined as $a_1 < 0.8$).

For some tests, the deformation is kinematically prescribed (e.g. a simple shear test with constant volume or a consolidated undrained triaxial test). Therefore, the calculation is quite simple for strain dependent constitutive models. For other element tests, only a part of the deformation is prescribed. In the case of a drained triaxial test, the axial stress is prescribed and the radial stresses remain constant, what causes an unknown deformation in radial direction. A nonlinear system of equations has to be solved for such a mixed problem.

In the case of the principal stress rotation, the stress is prescribed. Therefore the unknowns in the calculation are the time derivatives of the deformation $\dot{a}_1(t)$, $\dot{a}_2(t)$, $\dot{a}_3(t)$ and $\dot{a}_4(t)$. They can be merged in the vector $\dot{\boldsymbol{a}}$

$$\dot{\boldsymbol{a}}(t) = \begin{bmatrix} \dot{a}_1(t) \\ \dot{a}_2(t) \\ \dot{a}_3(t) \\ \dot{a}_4(t) \end{bmatrix} \quad . \tag{5.24}$$

The stress rate $\dot{\boldsymbol{T}}(t)$ is prescribed in stress controlled simulations for a given time step Δt. For constitutive relations, which are strain dependent formulated $\overset{\circ}{\boldsymbol{T}} = \boldsymbol{h}(\boldsymbol{T}, \boldsymbol{D}, \dots)$ (e.g. Hypoplasticity and Barodesy) with the given actual stress state $T(t)$, the deformation $a(t)$ and its estimated time derivative $\dot{a}$ yield the stress rate $\dot{T}(\dot{a})$. The vector $\dot{a}(t)$ has to be determined with Newton's method.

Also for stress dependant constitutive models $D = f(\mathring{T}, T, \ldots)$ (e.g. Sanisand) Newton's method is required. Since the constitutive relation is calculated with an objective stress rate $\mathring{T}$, the stress rate $\dot{T}$ has to be calculated from the resulting stretching and in general $\dot{T} \neq \mathring{T}$ holds true for non-rectilinear deformations.

5.4.1   Newton's method for Hypoplasticity and Barodesy

Since the stress state history $T(t)$ is prescribed, also the stress rate $\dot{T}(t)$ is known. For the following calculations the time integration is discretized for each time step Δt from t_i to $t_i + \Delta t$ as follows

$$\dot{T}(t_i) = \frac{T(t_i + \Delta t) - T(t_i)}{\Delta t} = \frac{T_{i+1} - T_i}{\Delta t} = \dot{T}_i \quad . \tag{5.25}$$

To achieve the correct stress rate, the corresponding derivative of the deformation $\dot{a}$ has to be found. For this task, Newton's method is employed. In the following, each iteration step is denoted with the superscript (n) and each time step is denoted with the subscript i. In the first iteration $n = 0$ the value of $\dot{a}_i^{(n)}$ is estimated. With this estimation, the stretching $D\left(\dot{a}_i^{(n)}\right)$ and the spin $W\left(\dot{a}_i^{(n)}\right)$ are calculated. At the begin of the calculation is $a(t = 0) = 0$. The stress rate $\dot{T}_i^{(n)}$ is calculated by

$$\dot{T}_i^{(n)} = \mathring{T}_i\left(T_i, D\left(\dot{a}_i^{(n)}\right), E_i\right) + W\left(\dot{a}_i^{(n)}\right) T_i - T_i W\left(\dot{a}_i^{(n)}\right) \quad , \tag{5.26}$$

where E_i denotes the vector of internal variables at the beginning of the time step, e.g. the void ratio e_i or the intergranular strain δ_i. The residual $R_{\mathrm{N}i}^{(n)}$ is the difference between the prescribed stress rate $\dot{T}_i$ and the obtained stress rate $\dot{T}_i^{(n)}\left(D\left(\dot{a}_i^{(n)}\right), W\left(\dot{a}_i^{(n)}\right), T_i, E_i\right)$ in the nth iteration of the ith time step. The new estimation of $\dot{a}_i^{(n+1)}$ is calculated by

$$\dot{a}_i^{(n+1)} = \dot{a}_i^{(n)} - \left(\frac{\partial \dot{T}_i}{\partial \dot{a}_i}\left(\dot{a}_i^{(n)}\right)\right)^{-1} R_{\mathrm{N}i}^{(n)} \quad . \tag{5.27}$$

If the norm of the residual $R_{\mathrm{N}i}^{(n)}$ is smaller than a given tolerance, $\dot{a}_i^{(n)}$ is accepted and a time integration with this $\dot{a}_i$ is performed

$$a_{i+1} = a(t_{i+1}) = a(t_i) + \dot{a}_i \Delta t = a_i + \dot{a}_i \Delta t \quad . \tag{5.28}$$

The time integration of the stress T_{i+1} and the internal variables E_{i+1} follow in a similar way.

5.4.2   Newton's method for Sanisand

As already mentioned, Sanisand has a different formulation. Therefore, Newton's method has to be adopted. Here the objective stress rate $\overset{\circ}{T}_i^{(n)}$ is estimated in the first iteration step. This results in an stretching tensor $D_i^{(n)} = f\left(\overset{\circ}{T}_i^{(n)}, T_i, E_i\right)$ from which the components of $\dot{a}_i^{(n)}$ can be calculated with

$$\dot{a}_{3,i}^{(n)} = D_{11}^{(n)}(1 + a_{3,i}) \quad , \tag{5.29}$$

$$\dot{a}_{2,i}^{(n)} = D_{22}^{(n)}(1 + a_{2,i}) \quad , \tag{5.30}$$

$$\dot{a}_{4,i}^{(n)} = D_{33}^{(n)}(1 + a_{4,i}) \quad \text{and} \tag{5.31}$$

$$\dot{a}_{1,i}^{(n)} = \frac{D_{12}^{(n)}(1 + a_{2,i})(1 + a_{3,i}) + \dot{a}_{3,i}^{(n)} a_{1,i}}{1 + a_{3,i}} \quad . \tag{5.32}$$

With the known $\dot{a}_i^{(n)} = \dot{a}\left(\overset{\circ}{T}_i^{(n)}\right)$ the spin tensor $W_i^{(n)} = W\left(\dot{a}_i^{(n)}\right) = W\left(\dot{a}\left(\overset{\circ}{T}_i^{(n)}\right)\right)$ and the stress rate $\dot{T}_i^{(n)}$ can be calculated

$$\dot{T}_i^{(n)} = \overset{\circ}{T}_i + W\left(\dot{a}\left(\overset{\circ}{T}_i^{(n)}\right)\right) T_i - T_i W\left(\dot{a}\left(\overset{\circ}{T}_i^{(n)}\right)\right) \quad . \tag{5.33}$$

The new estimation of $\overset{\circ}{T}_i^{(n+1)}$ is calculated with

$$\overset{\circ}{T}_i^{(n+1)} = \overset{\circ}{T}_i^{(n)} - \left(\frac{\partial \dot{T}_i}{\partial \overset{\circ}{T}_i}\left(\overset{\circ}{T}_i^{(n)}\right)\right)^{-1} R_{\mathrm{N}_i}^{(n)} \quad . \tag{5.34}$$

Again, if the norm of the residual $R_{\mathrm{N}_i}^{(n)} = \dot{T}_i^{(n)} - \dot{T}_i$ is smaller than a given tolerance, $\overset{\circ}{T}_i^{(n)}$ is accepted as solution and a time integration step for stress T, deformation a and internal variables E is executed. The integration of stretching has to be handled differently and is described in the following.

5.4.3   Integration of the rate of deformation

Since the deformation also contains a rotation, the time integration of the rate of deformation D in the form $\varepsilon_{i+1} = \varepsilon_i + D_i \Delta t$ is not correct. To get the correct logarithmic strain (also known as true or Hencky strain) in the reference configuration the stretching has to be rotated back into the reference configuration.

To find the correct rotation matrix a polar decomposition of the deformation gradient $\boldsymbol{F}$ has to be made. The deformation gradient can also be written as

$$\boldsymbol{F} = \boldsymbol{QU} \quad , \tag{5.35}$$

with the rotation matrix $\boldsymbol{Q}$ and the right stretch tensor $\boldsymbol{U}$. The rotation matrix $\boldsymbol{Q}$ is an orthonormal matrix, i.e. $\boldsymbol{QQ}^\mathsf{T} = \boldsymbol{Q}^\mathsf{T}\boldsymbol{Q} = \boldsymbol{I}$ and $\boldsymbol{Q}^{-1} = \boldsymbol{Q}^\mathsf{T}$. For the calculation of $\boldsymbol{Q}$, $\boldsymbol{U}$ has to be determined. Therefore, the deformation gradient $\boldsymbol{F}$ is multiplied by its transposed

$$\boldsymbol{F}^\mathsf{T}\boldsymbol{F} = (\boldsymbol{QU})^\mathsf{T}(\boldsymbol{QU}) = \boldsymbol{U}^\mathsf{T}\boldsymbol{Q}^\mathsf{T}\boldsymbol{QU} = \boldsymbol{U}^\mathsf{T}\boldsymbol{U} = \boldsymbol{B} \quad . \tag{5.36}$$

This result is called the right Cauchy-Green deformation tensor $\boldsymbol{B}$. In this matrix the rotational part $\boldsymbol{Q}$ of $\boldsymbol{F}$ is eliminated. In a next step $\boldsymbol{B}$ has to be rotated in such a way that a diagonal matrix is found to get the square root of $\boldsymbol{B}'$ (note, the new coordinate system is denoted with a $'$), since

$$\boldsymbol{B}' = \boldsymbol{U}'^2 \rightsquigarrow \boldsymbol{U}' = \sqrt{\boldsymbol{U}'^2} \quad . \tag{5.37}$$

Finding a diagonal matrix is done calculating the eigenvalues λ_i (which are the elements in the diagonal matrix). For the further calculation, additionally the right eigenvectors v_i of the matrix (which build the rotation matrix $\boldsymbol{V} = [v_1, v_2, v_3]$) are needed too. To get $\boldsymbol{U}$ in the reference system, $\boldsymbol{U}'$ has to be rotated back after taking the square root

$$\boldsymbol{U} = \boldsymbol{V}\boldsymbol{U}'\boldsymbol{V}^{-1} \quad . \tag{5.38}$$

With the known right stretching tensor $\boldsymbol{U}$ the rotation matrix $\boldsymbol{Q}$ can calculated from the deformation gradient

$$\boldsymbol{Q} = \boldsymbol{F}\boldsymbol{U}^{-1} \quad . \tag{5.39}$$

The true strain can then be calculated with the equation

$$\varepsilon_{i+1} = \varepsilon_i + \boldsymbol{Q}_i^\mathsf{T}\boldsymbol{D}_i\boldsymbol{Q}_i\Delta t \quad . \tag{5.40}$$

5.5   Tests with a $1\gamma2\varepsilon$ apparatus

In this section, some results of tests with the $1\gamma2\varepsilon$ apparatus [58] (schematic sketch in Fig. 5.10) are compared with numerical simulations with the above chosen constitutive models.

5.5.1   Testing device

The apparatus consists of a parallelogram (OABC), which encloses the sample. The parallelogram can undergo horizontal and vertical elongations and a shear deformation (tilting of sides AB and CO). The sides AO and BC remain horizontal and the sides AB and CO remain parallel during the deformation. To achieve this deformation, four motors are necessary, which control the elongations. The point O is fixed. Each side consists of six comb-shaped segments, interlocked in such a way that they maintain the continuity of the side. To provide a homogeneous strain the relative deformation of the segments is equal disturbed.

The four sides are linked with hinges in such a way that they can be tilted. Shear strain is imposed by applying a horizontal displacement at the centre of AB by a fifth motor, so that the side tilts around A and, correspondingly, the side CO around O.

The corners B, C and O are equipped with strain-gauged hinges consisting of two complete strain-gauge bridges placed inside a cylinder. This allows measurement of the forces applied at each corner in horizontal and vertical directions, Fig. 5.10. Assuming homogeneous stress in the sample the stress state can be determined from the applied forces.

The deformation is controlled by the velocities imposed by the motors. Strain paths can be easily prescribed, since the deformation is controlled. In our case, we have stress path control and this requires the use of servo-control, where the deformation is adjusted according to the measured forces.

Schneebeli cylinders of polyvinyl chloride (PVC) with different diameters (1.5, 3 and 3.5 mm) are used in these tests, which can be seen as a two dimensional analogous material to soil [102].

5.5.2   Stress path

The complex cyclic shear tests with constant principal stress values [59] are numerically simulated.

The sample is initially loaded with a vertical stress $T_{22} = T_1 = -90$ kPa and lateral stress $T_{11} = T_3 = -50$ kPa. T_1, T_2 and T_3 denote the maximum, intermediate and minimum principal stresses, respectively. The sample is deformed in a way that the directions of the principal stresses rotate while their values remain constant. This stress evolution can be described by the

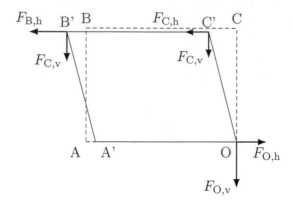

Figure 5.10: Deformed configuration of the $1\gamma2\varepsilon$ apparatus with force measurements

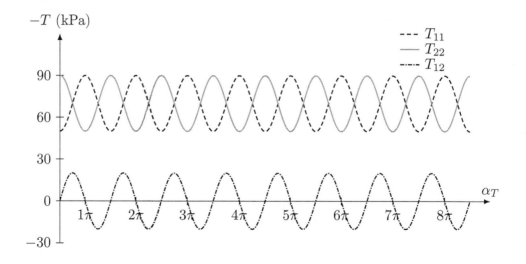

Figure 5.11: Prescribed stresses in the $1\gamma2\varepsilon$ apparatus

following equations

$$T_{11} = \frac{T_1 + T_3}{2} + \frac{T_1 - T_3}{2} \cos 2\alpha_T,$$

$$T_{22} = \frac{T_1 + T_3}{2} - \frac{T_1 - T_3}{2} \cos 2\alpha_T \quad \text{and} \qquad (5.41)$$

$$T_{12} = -\frac{T_1 - T_3}{2} \sin 2\alpha_T$$

(cf. Figure 5.11). The angle of the rotation of the stresses is denoted with the direction of the principal stress α_T. Fig. 5.12a shows several material points for one cycle of α_T. In the upper area the stresses acting on the elements in

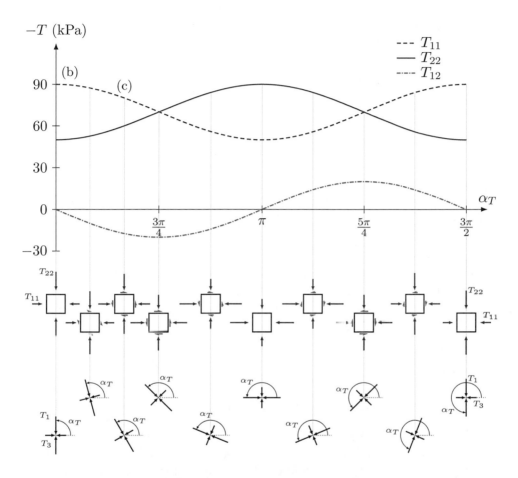

(a) Rotation of stress visualised on elements

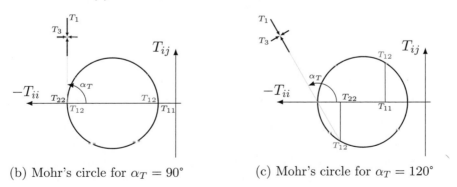

(b) Mohr's circle for $\alpha_T = 90°$

(c) Mohr's circle for $\alpha_T = 120°$

Figure 5.12: Principal stress rotation visualised with elements and Mohr's circles

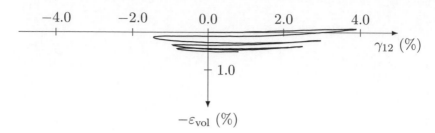

Figure 5.13: Volumetric strain vs. shear strain, experimental results [59]

the original coordinate system are drawn. The increase and decrease of the shear and normal stresses can be seen. In the lower part, the direction of the principal stresses (α_T) is shown. Here the monotonic increase of the angle α_T can be seen. In Fig. 5.12b and 5.12c Mohr's circles are drawn for $\alpha_T = 90°$ and for $\alpha_T = 120°$. Note, it can be seen that the directions of the principal stresses T_1 and T_3 are variable, whereas the direction of T_2 is fixed and coincides with the x_3-direction (out of the plane).

The numerical simulations are conducted as element tests. An assumption has to be made for the boundary condition in the out of plane direction (x_3 in Fig. 5.9). The out of plane stress has to be constant to keep the invariants of the stress tensor constant. Starting from an isotropic stress state (mean effective stress $-p = T_1 = T_2 = T_3 = -30\,\text{kPa}$) the stress is anisotropically increased to $T_1 = -90\,\text{kPa}$ and $T_2 = T_3 = -50\,\text{kPa}$.

The finite element program Plaxis is used for the simulation with Hardening Soil, because this model is not explicitly published. All other calculations are conducted with Matlab. The codes for the calculation can be found in Appendix E.

5.5.3   Results

The volumetric strain ($\varepsilon_{\text{vol}} = \varepsilon_{11} + \varepsilon_{22}$) of the laboratory test is plotted versus the shear strain ($\gamma_{12} = 2\varepsilon_{12}$), Fig. 5.13. An overall decrease of the volume can be seen in the laboratory test.

The numerical results of the two elastoplastic constitutive models (Sanisand and Hardening Soil) are shown in Figure 5.14. The Hardening Soil Small Model (Fig. 5.14a) does not show any volumetric response. Contrary, Sanisand is able to reproduce the volumetric compaction qualitatively (Fig. 5.14b). The hypoplastic and barodetic models are also able to reproduce a compaction. None of the models is able to reproduce the dilatation at the start of the test and at the end of each cycle.

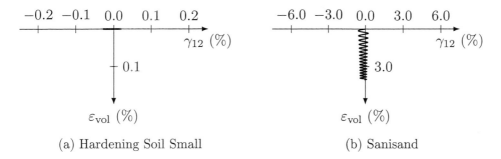

(a) Hardening Soil Small   (b) Sanisand

Figure 5.14: Results of the simulations with the elastoplastic constitutive models

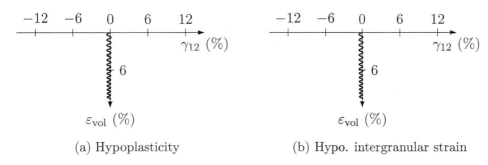

(a) Hypoplasticity   (b) **Hypo.** intergranular strain

Figure 5.15: Volumetric behaviour of **Hyp**oplastic models

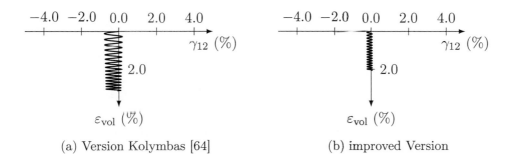

(a) Version Kolymbas [64]   (b) improved Version

Figure 5.16: Volumetric behaviour of Barodesy

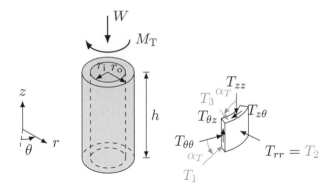

Figure 5.17: Schematic sketch of a hollow cylinder test

5.6   Tests with a hollow cylinder apparatus

A rotation of principal stresses can also be applied in a hollow cylinder test
(e.g. [11, 55, 83, 117, 129]). A schematic sketch of a hollow cylinder is shown
in Fig. 5.17. The soil sample is loaded with an axial force W, an inner pressure
p_i, an outer pressure p_o and a torque M_T.

It is not possible to measure the stress in the sample directly and the stress
is not constant distributed in the specimen, e.g. for different inner and outer
pressures the radial stress changes from p_i to p_o along the wall thickness. To
compare the results of the laboratory test, averaged stresses are used. Hence,
some assumptions about the stress distribution and the material behaviour
have to be made. In the literature several solutions can be found, depending
on whether the averaging of the stress is made over the thickness of the wall
(e.g. Hight *et al.* [54], Miura *et al.* [83] and Yang *et al.* [129]) or the volume
(e.g. Naughton and O'Kelly [86] and Sayao and Vaid [99]), if it is assumed
that the cylinder is thin-walled (Yang *et al.* [129]) or thick-walled (Naughton
and O'Kelly [86]), or if the stresses are calculated for a linear elastic material
(Hight *et al.* [54]), partially plastic ([99]) or an average of elastic and plastic
solutions (Yang *et al.* [129]). Wijewickreme [128] has investigated the influence
of a non-linear elasticity in soil behaviour with finite element calculations in his
PhD thesis. He could not find a significantly influence on the average stress.
In general, the differences of the averaged stresses are small for all different
assumptions. The different equations for the average stresses can be seen in
Tab. 5.6. In Tab. 5.7, the equations for the strains are listed. Here, u_o, u_i and
z denote outer radial, inner radial and vertical displacement, respectively and
the rotation of the top plate is denoted with ω.

A drained test from Tong *et al.* [117] has been simulated with Sanisand, Hypoplasticity and Barodesy. No simulations have been performed with Hard-

Table 5.6: Calculation of the stress in a hollow cylinder

	T_{zz}	$T_{\theta\theta}$	T_{rr}	$T_{z\theta}$
Hight et al. [54]	$\dfrac{W}{\pi(r_o^2 - r_i^2)} + \dfrac{p_o r_o^2 - p_i r_i^2}{r_o^2 - r_i^2}$	$\dfrac{p_o r_o - p_i r_i}{r_o - r_i}$	$\dfrac{p_o r_o + p_i r_i}{r_o + r_i}$	$\dfrac{3 M_T}{2\pi(r_o^3 - r_i^3)}$
Miura et al. [83]	$\dfrac{W + p_o\pi(r_o^2 - r_i^2) - p_i\pi r_i^2}{\pi(r_o^2 - r_i^2)}$	$\dfrac{p_o(4r_o^2 + r_o r_i + r_i^2) - p_i(r_o^2 + r_o r_i + 4r_i^2)}{3(r_o^2 - r_i^2)}$	$\dfrac{p_o(2r_o^2 - r_o r_i - r_i^2) + p_i(r_o^2 + r_o r_i - 2r_i^2)}{3(r_o^2 - r_i^2)}$	$\dfrac{3 M_T}{2\pi(r_o^3 - r_i^3)}$
Naughton and O'Kelly [36]	$\dfrac{W}{\pi(r_o^2 - r_i^2)} + \dfrac{p_o r_o^2 - p_i r_i^2}{r_o^2 - r_i^2}$	$\dfrac{p_o r_o^2 - p_i r_i^2}{r_o^2 - r_i^2} + \dfrac{2 r_i^2 r_o^2 (p_o - p_i)\ln(r_o/r_i)}{(r_o^2 - r_i^2)^2}$	$\dfrac{p_o r_o^2 - p_i r_i^2}{r_o^2 - r_i^2} - \dfrac{2 r_i^2 r_o^2 (p_o - p_i)\ln(r_o/r_i)}{(r_o^2 - r_i^2)^2}$	$\dfrac{4 M_T (r_o^3 - r_i^3)}{3\pi(r_o^4 - r_i^4)(r_o^2 - r_i^2)}$
Sayao and Vaid [99]	$\dfrac{W}{\pi(r_o^2 - r_i^2)} + \dfrac{p_o r_o^2 - p_i r_i^2}{r_o^2 - r_i^2}$	$\dfrac{p_o r_o - p_i r_i}{r_o - r_i}$	$\dfrac{p_o r_o + p_i r_i}{r_o + r_i}$	$\dfrac{4 M_T (r_o^3 - r_i^3)}{3\pi(r_o^2 - r_i^2)(r_o^4 - r_i^4)}$
Yang et al. [129]	$\dfrac{W}{\pi(r_o^2 - r_i^2)} + \dfrac{p_o(r_o^2 - r_d^2) - p_i r_i^2}{r_o^2 - r_i^2}$	$\dfrac{p_o r_o - p_i r_i}{r_o - r_i}$	$\dfrac{p_o r_o + p_i r_i}{r_o + r_i}$	$\dfrac{M_T}{2}\left[\dfrac{3}{2\pi(r_o^3 - r_i^3)} + \dfrac{4(r_o^3 - r_i^3)}{3\pi(r_o^2 - r_i^2)(r_o^4 - r_i^4)}\right]$

Table 5.7: Calculation of the strain in a hollow cylinder

	ε_{zz}	$\varepsilon_{\theta\theta}$	ε_{rr}	$\varepsilon_{z\theta}$
Hight et al. [54]	$\dfrac{z}{h}$	$\dfrac{u_o + u_i}{r_o + r_i}$	$\dfrac{u_c - u_i}{r_o - r_i}$	$\dfrac{\omega(r_o^3 - r_i^3)}{3h(r_o^2 - r_i^2)}$
Miura et al. [83]	$\dfrac{z}{h}$	$\dfrac{u_o + u_i}{r_o + r_i}$	$\dfrac{u_o - u_i}{r_o - r_i}$	$\dfrac{\omega(r_o^3 - r_i^3)}{3h(r_o^3 - r_i^3)}$
thick-walled [86]	$\dfrac{z}{h}$	$\dfrac{u_o + u_i}{r_o + r_i}$	$\dfrac{u_o - u_i}{r_o - r_i}$	$\dfrac{\omega(r_o^3 - r_i^3)}{3h(r_o^3 - r_i^3)}$
Sayao and Vaid [99]	$\dfrac{z}{h}$	$\dfrac{u_o + u_i}{r_o + r_i}$	$\dfrac{u_o - u_i}{r_o - r_i}$	$\dfrac{\omega(r_o^3 - r_i^3)}{3h(r_o^3 - r_i^3)}$
Yang et al. [129]	$\dfrac{z}{h}$	$\dfrac{u_o + u_i}{r_o + r_i}$	$\dfrac{u_o - u_i}{r_o - r_i}$	$\dfrac{\omega(r_o^3 - r_i^3)}{3h(r_o^2 - r_i^2)}$

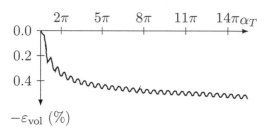

Figure 5.18: Volumetric strain vs. number of cycles in a hollow cylinder test [117]

ening Soil Small, since this model shows no volumetric reaction in Section 5.5, and Hypoplasticity with intergranular strain, since there are just very small differences between the two hypoplastic models. The stresses are calculated with the assumptions of a thin walled cylinder, linear elastic material behaviour and for the shear stress the average of the elastic and the plastic solution is used according to Yang *et al.* [129].

The initial conditions for the test series III in [117] are given as void ratio $e = 0.737 \pm 0.007$ $(D_r \approx (70 \pm 2)\,\%)$ and isotropic stress $p = 30\,\text{kPa}$. The sample is then anisotropically compressed so that $T_1 = -157.4\,\text{kPa}$, $T_2 = -100\,\text{kPa}$ and $T_3 = -42.6\,\text{kPa}$. This results in a mean normal stress $p = 100\,\text{kPa}$, a deviatoric stress $q = 99.4\,\text{kPa}$ and $b_T = 0.5$. Here b_T denotes the intermediate principal stress parameter

$$b_T = \frac{T_2 - T_3}{T_1 - T_3} \quad , \tag{5.42}$$

i.e. for triaxial compression $b_T = 0$ and for triaxial extension $b_T = 1$. Note, the deviatoric stress q is in this general case not the difference between the maximum and the minimum principal stress, but

$$q = \sqrt{\frac{(T_1 - T_2)^2 + (T_2 - T_3)^2 + (T_3 - T_1)^2}{2}} \quad . \tag{5.43}$$

The result of the laboratory test is shown in Fig. 5.18.

The results of the simulations are shown in Fig. 5.19, 5.20, 5.21 and 5.22, where the evolution of volumetric strain ε_{vol} and a measure for deviatoric strain $\varepsilon_{\text{q}} = \sqrt{\frac{3}{2}\,\text{tr}\,\varepsilon^{*2}}$ are shown, ε^* is here the deviatoric part of the strain tensor ε as defined in (5.4). All simulations show dilatation, which is opposite to the behaviour of the physical test. The evolutions of the deviatoric strains predicted by the models differ considerably. It is also conceivable that the elastic solution used for the evaluation of the physical tests is not realistic.

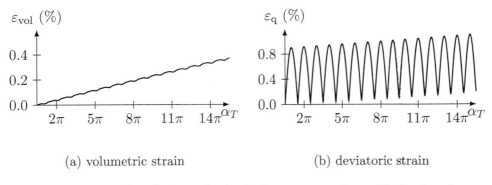

(a) volumetric strain   (b) deviatoric strain

Figure 5.19: Simulation of principal stress rotation with Sanisand

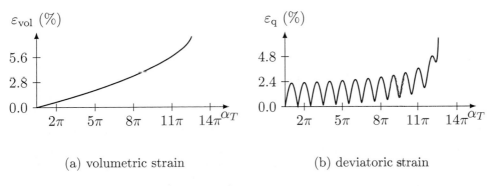

(a) volumetric strain   (b) deviatoric strain

Figure 5.20: Simulation of principal stress rotation with Hypoplasticity

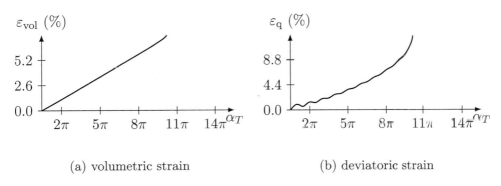

(a) volumetric strain   (b) deviatoric strain

Figure 5.21: Simulation of principal stress rotation with Barodesy in the version of Kolymbas [64]

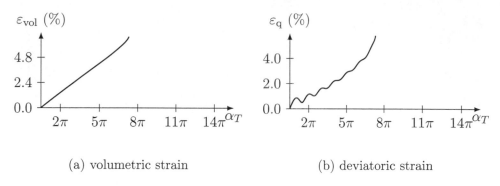

(a) volumetric strain   (b) deviatoric strain

Figure 5.22: Simulation of principal stress rotation with the improved version of Barodesy

5.7   Conclusions

It can be seen from the numerical simulations, that constitutive models, which use only the principal stresses can, clearly, not reproduce the volumetric changes of soil due to principal stress rotation. This is shown here for the Hardening Soil model, but applies also for the often-used simpler elastoplastic model with a Mohr-Coulomb failure criterion. Therefore, problems can arise, if such an inappropriate model is used for calculations with rotations of the principal stresses. In case of elastoplastic models this shortcoming can be eliminated with a reformulation of the failure surface and flow rule in a six dimensional stress space as shown in [55]. The hypoplastic and barodetic models are able to reproduce a volumetric change without any further modification.

Advanced constitutive relations can model a volumetric change of soil due to a rotation of the principal stresses. However, as here shown there are cases, where these models predict a dilatant behaviour for this type of loading whereas the corresponding laboratory test shows contractant behaviour. Note, that the used constitutive relations model the volumetric behaviour of drained triaxial and simple shear tests quite satisfactory and similar to each other. This discrepancy in the results of conventional laboratory tests and principal stress rotation experiments is a clear hint that further research is needed to obtain a reliable volumetric behaviour of constitutive models under complex stress paths.

Chapter 6

Stability of infinite slopes

In this chapter, the stability of infinite slopes is investigated with different models. The mechanical model of the infinite slope seems very simple at a first glance and is hence often implemented into Geographical Information Systems (e.g. [80, 81, 82]). This model is in fact very clear and can train the ability to interpret slope stability problems mechanically.

6.1 Friction angle and angle of dilatancy of soil

In this chapter, the terms friction angle and angle of dilatancy are extensively used and also further definitions of friction angles are introduced. In order to make the difference clearer, the terms already used are explained again. The mobilized friction angle is calculated with, T_1 and T_3, the major and the minor principal stress, respectively,

$$\sin \varphi_{\mathrm{mob}} = \frac{T_3 - T_1}{T_1 + T_3} \tag{6.1}$$

and is related to the Mohr-Coulomb failure criterion. This mobilized friction angle reaches a maximum value during shearing, the so-called peak friction angle $\varphi_{\mathrm{p}} = \max \varphi_{\mathrm{mob}}$, and (for initially dense samples) will subsequently decrease to the critical friction angle φ_{c}. The specimen will change its volume throughout shearing. An angle of dilatancy can be calculated to be used in a flow rule of an elastoplastic material model. For example, in triaxial conditions, it follows for a Mohr-Coulomb type flow rule:

$$\tan \psi_{\mathrm{mob}} = -\frac{\operatorname{tr} \boldsymbol{D}}{2D_1 - \operatorname{tr} \boldsymbol{D}} \quad, \tag{6.2}$$

where we use the same index mob as for the mobilized friction angle to emphasize that these two values are coupled. For elastoplastic models

$$\boldsymbol{D} = \boldsymbol{D}^{\mathrm{e}} + \boldsymbol{D}^{\mathrm{p}} \tag{6.3}$$

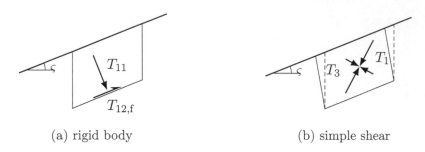

(a) rigid body   (b) simple shear

Figure 6.1: Infinite slope, possible failure of an element.

holds true, where $\boldsymbol{D}^{\mathrm{e}}$ and $\boldsymbol{D}^{\mathrm{p}}$ are the elastic and the plastic parts of the stretching, respectively. For a sample after a long constant yielding $\boldsymbol{D}^{\mathrm{e}}$ gradually vanishes and (6.2) yields

$$\tan \psi = \frac{\operatorname{tr} \boldsymbol{D}^{\mathrm{p}}}{2D_1^{\mathrm{p}} - \operatorname{tr} \boldsymbol{D}^{\mathrm{p}}} \tag{6.4}$$

what is the same as (1.5). We denote the value of ψ_{mob} at the peak of shear strength with ψ_{p}. At critical state $\psi_{\mathrm{mob}} = \psi_c = 0$.

Another mobilized friction angle can be calculated from a simple shear or direct shear test with the applied normal stress T_{22} and the measured shear stress T_{12}

$$\tan \phi_{\mathrm{mob}} = \frac{T_{12}}{T_{22}} \quad , \tag{6.5}$$

which is related to a Coulomb failure criterion. Again, we find a peak value for dense specimens $\phi_{\mathrm{p}} = \max \phi_{\mathrm{mob}}$ and a critical value ϕ_c for large shear strains. Only the peak friction angle is used in the following calculations. Hence, we will abbreviate to $\varphi := \varphi_{\mathrm{p}}$ and $\psi := \psi_{\mathrm{p}}$ in triaxial tests, as well as $\phi := \phi_{\mathrm{p}}$ in simple or direct shear tests.

6.2   Mohr-Coulomb – elastoplastic

The standard derivation of the stability of an infinite slope relates the normal stress T_{22} at the bottom of a lamella (Fig. 6.1a) with the shear stress at failure $T_{12,f}$ using a Coulomb failure criterion. With the tacit assumption of $\varphi_{\mathrm{p}} = \phi_{\mathrm{p}}$ it follows that the inclination of the slope ς at the limit state is equal to the friction angle φ (e.g. Kolymbas *et al.* [65]). Teunissen and Spierenburg [114] posed the question as to whether the material strength in the lamella is high enough to remain as a rigid body in this limit state calculation. They employed an ideal plastic material model with a Mohr-Coulomb failure criterion and non-associated flow rule and found out that $\varphi = \phi$ is only valid for an associated flow rule, which is in line with the collapse theorems of plasticity theory (Goldscheider [40]). However, soil does not have a dilatancy angle equal to the friction angle. In triaxial tests, the dilatancy of Hostun Sand at peak is between $\varphi_p/4$ and $\varphi_p/3$ [22]. Teunissen and Spierenburg [114] proposed a relation for the limit state of a slope, which depends on the friction angle and the dilatancy angle

$$\tan \varsigma = \frac{\sin \varphi \cos \psi}{1 - \sin \varphi \sin \psi} \quad . \tag{6.6}$$

If the slope failure is assumed to be a simple shear mechanism (Fig. 6.1b), equation (6.6) can be derived analytically from a simple shear element test employing a linear elastic-perfectly plastic material model with a Mohr-Coulomb yield function and a non-associated flow rule.

The Mohr-Coulomb failure criterion for a material without cohesion is

$$f_y(\boldsymbol{T}) = \frac{T_1 + T_3}{2} \sin \varphi - \frac{T_1 - T_3}{2} \quad . \tag{6.7}$$

For plain strain conditions the principal stresses are

$$T_{1,3} = \frac{T_{11} + T_{22}}{2} \pm \sqrt{\frac{(T_{11} - T_{22})^2}{4} + T_{12}^2} = T_m \pm T_q \quad , \tag{6.8}$$

substituting (6.8) into (6.7) yields

$$f_y = T_m \sin \varphi + T_q = \frac{T_{11} + T_{22}}{2} \sin \varphi + \sqrt{\frac{(T_{11} - T_{22})^2}{4} + T_{12}^2} \quad . \tag{6.9}$$

The plastic potential g_y for Mohr-Coulomb is defined as

$$g_y = T_m \sin \psi + T_q \quad . \tag{6.10}$$

In the following, a modified Voigt notation[1] will be used for calculations with elastoplastic models. This means that the stress and strain tensors are represented as

$$\boldsymbol{T} = \begin{bmatrix} T_{11} \\ T_{22} \\ T_{12} \\ T_{33} \end{bmatrix} \quad \text{and} \quad \boldsymbol{D} = \begin{bmatrix} D_{11} \\ D_{22} \\ 2D_{12} \\ D_{33} \end{bmatrix}$$

in the plane strain condition, respectively. The factor 2 in the stretching guarantees that the Helmholtz free energy is the result of a multiplication of

---

[1] The original Voigt notation of a second order tensor in the $\mathbb{R}^3$ is written as $\boldsymbol{T} = \begin{bmatrix} T_{11} & T_{22} & T_{33} & T_{23} & T_{13} & T_{12} \end{bmatrix}^\mathsf{T}$

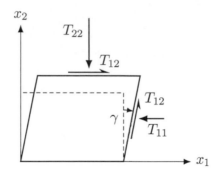

Figure 6.2: Schematic simple shear test

T and ε in the Voigt notation. For the calculation in the elastic region also the elastic stiffness tensor C^{e} is required, which is for plain strain

$$C^{\mathrm{e}} = \frac{E}{(1+\nu)(1-2\nu)} \begin{bmatrix} 1-\nu & \nu & 0 & \nu \\ \nu & 1-\nu & 0 & \nu \\ 0 & 0 & 0.5-\nu & 0 \\ \nu & \nu & 0 & 1-\nu \end{bmatrix} . \tag{6.11}$$

In the elastic region the stress rate is calculated by

$$\dot{T} = C^{\mathrm{e}} D = C^{\mathrm{e}} D^{\mathrm{e}} \tag{6.12}$$

When the yield surface is reached the elastoplastic material tensor C^{ep} has to be used, which depends on the actual stress state and is calculated with

$$C^{\mathrm{ep}} = C^{\mathrm{e}} - \frac{C^{\mathrm{e}} m n^{\mathsf{T}} C^{\mathrm{e}}}{n^{\mathsf{T}} C^{\mathrm{e}} m} . \tag{6.13}$$

Herein, n and m are the normal vectors on the yield surface f_{y} and the plastic potential g_{y} respectively. For the Mohr-Coulomb failure criterion and plane strain, n reads

$$n = \frac{\partial f_{\mathrm{y}}}{\partial T} = \begin{bmatrix} \frac{\partial f_{\mathrm{y}}}{\partial T_{11}} \\ \frac{\partial f_{\mathrm{y}}}{\partial T_{22}} \\ \frac{\partial f_{\mathrm{y}}}{\partial T_{12}} \\ \frac{\partial f_{\mathrm{y}}}{\partial T_{33}} \end{bmatrix} = \begin{bmatrix} \frac{T_{11}-T_{22}}{2T_{\mathrm{q}}} + \sin\varphi \\ -\frac{T_{11}-T_{22}}{2T_{\mathrm{q}}} + \sin\varphi \\ \frac{2T_{12}}{T_{\mathrm{q}}} \\ 0 \end{bmatrix} . \tag{6.14}$$

The normal on the plastic potential is determined as

$$m = \frac{\partial g_{\mathrm{y}}}{\partial T} = \begin{bmatrix} \frac{\partial g_{\mathrm{y}}}{\partial T_{11}} \\ \frac{\partial g_{\mathrm{y}}}{\partial T_{22}} \\ \frac{\partial g_{\mathrm{y}}}{\partial T_{12}} \\ \frac{\partial g_{\mathrm{y}}}{\partial T_{33}} \end{bmatrix} = \begin{bmatrix} \frac{T_{11}-T_{22}}{2T_{\mathrm{q}}} + \sin\psi \\ -\frac{T_{11}-T_{22}}{2T_{\mathrm{q}}} + \sin\psi \\ \frac{2T_{12}}{T_{\mathrm{q}}} \\ 0 \end{bmatrix} . \tag{6.15}$$

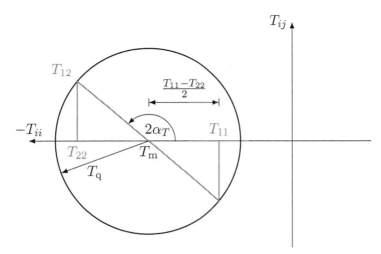

Figure 6.3: Mohr's circle

In the equations (6.14) and (6.15) the expression $\frac{T_{11}-T_{22}}{2T_q}$ can be substituted by $-\cos 2\alpha_T$ and $\frac{T_{12}}{T_q}$ by $\sin 2\alpha_T$ (see Fig. 6.3)

For the calculation of the maximum stress, a strain controlled simple shear test is computed. The shearing proceeds in the x_1-x_2-plane, with vertical loading in x_2-direction and constrained displacement in x_3-direction. The boundary conditions are hence as follows: zero normal strains in x_1- and x_3-directions and a constant normal stress in x_2-direction. These conditions result in the following system of equations when the yield surface is reached

$$\begin{bmatrix} \dot{T}_{11} =? \\ \dot{T}_{22} = 0 \\ \dot{T}_{12} =? \\ \dot{T}_{33} =? \end{bmatrix} = \boldsymbol{C}^{\mathrm{ep}} \begin{bmatrix} D_{11} = 0 \\ D_{22} =? \\ 2D_{12} \\ D_{33} = 0 \end{bmatrix} \quad . \tag{6.16}$$

The unknowns in these equations (indicated with question marks) can be calculated as

$$D_{22} = -\frac{D_{23}^{\mathrm{ep}}}{D_{22}^{\mathrm{ep}}} 2D_{12} \quad , \tag{6.17}$$

$$\dot{T}_{11} = \left(-\frac{D_{23}^{\mathrm{ep}} D_{12}^{\mathrm{ep}}}{D_{22}^{\mathrm{ep}}} + D_{13}^{\mathrm{ep}} \right) 2D_{12} \quad , \tag{6.18}$$

$$\dot{T}_{12} = \left(-\frac{D_{23}^{\mathrm{ep}} D_{32}^{\mathrm{ep}}}{D_{22}^{\mathrm{ep}}} + D_{33}^{\mathrm{ep}} \right) 2D_{12} \quad , \tag{6.19}$$

$$\dot{T}_{33} = \left(-\frac{D_{23}^{\mathrm{ep}} D_{42}^{\mathrm{ep}}}{D_{22}^{\mathrm{ep}}} + D_{43}^{\mathrm{ep}} \right) 2D_{12} \quad . \tag{6.20}$$

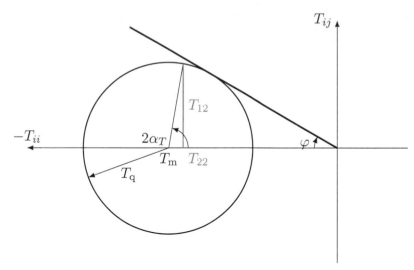

Figure 6.4: Mohr's diagram for the failure state

The final stress state is reached, when all stress rates are zero. From (6.18), (6.19) and (6.20) it can be seen that this is the case if the terms in parentheses vanish. Some manipulation yields for (6.18) and (6.20)

$$(\sin\psi - \cos 2\alpha_T)\sin 2\alpha_T = 0 \quad . \tag{6.21}$$

The normal stress rates $\dot{T}_{11}$ and $\dot{T}_{33}$ vanish for $\alpha_T = 0°$ or $\alpha_T = 90°$ ($\sin 2\alpha_T = 0$) and for $\alpha_T = 45° - \psi/2$ ($\sin\psi - \cos 2\alpha_T = 0$). The term in parentheses of (6.19) is

$$\cos^2 2\alpha_T - \cos 2\alpha_T(\sin\varphi + \sin\psi) + \sin\varphi\sin\psi = 0 \quad . \tag{6.22}$$

The solution of this quadratic equation is $\alpha_T = 45° - \psi/2$ and $\alpha_T = 45° - \varphi/2$.

It can be seen that for the case of $\alpha_T = 45° - \psi/2$ all stress rates are zero and the final stress state is reached. With the knowledge of this angle and the normal stress T_{22} the final shear stress can be calculated as (cf. Fig. 6.4),

$$T_{\mathrm{q}} = -T_{\mathrm{m}}\sin\varphi \quad , \tag{6.23}$$
$$T_{12} = T_{\mathrm{q}}\sin 2\alpha_T \quad , \tag{6.24}$$
$$T_{22} = T_{\mathrm{m}} - T_{\mathrm{q}}\cos 2\alpha_T \quad . \tag{6.25}$$

With the use of (6.23) in (6.25) and (6.24) the relation of T_{12} to T_{22} is

$$\frac{T_{12}}{T_{22}} = \frac{\sin\varphi\sin 2\alpha_T}{1 - \sin\varphi\cos 2\alpha_T} \quad . \tag{6.26}$$

The equation (6.6) results by substituting $2\alpha_T$ with $90° - \psi$.

The flow rule (as derived from a plastic potential with respect to the stress) in such models implies coaxiality of the stress tensor $\boldsymbol{T}$ and the plastic stretching tensor $\boldsymbol{D}^{\mathrm{p}}$. However, calculations with an elastoplastic material model with a non-coaxial flow rule (Yu [130], Yu and Yuan [131]) will predict the same values of ς. From a relation between Coulomb and Mohr-Coulomb friction angle proposed by Thornton and Zhang [116] for simple shear conditions in a discrete element computation, it follows for a non-coaxial flow rule (similar to a relationship suggested by Tatsuoka *et al.* [111]) that

$$\tan\varsigma = \frac{\sin\varphi\cos(\psi+2\iota)}{1-\sin\varphi\sin(\psi+2\iota)} \tag{6.27}$$

with $\iota = \alpha_T - \alpha_D$, being the angle of non-coaxiality, i.e. the deviation of the principal directions of the stress tensor α_T and the strain rate tensor α_D. Numerical simple shear experiments with discrete elements showed for $K_T = 0.5$ that the angle of non-coaxiality starts with $\iota < 0$ and monotonically increases to $\iota = 0$ (i.e. coaxiality) for large strains. This means that ς from (6.27) is smaller or equal to ς from (6.6), which confirms the computations with the elastoplastic model mentioned above [131]. For $K_T = 1$ the angle of non-coaxiality $\iota \approx 0$ applies throughout the entire shearing process [116].

Following the definition from Fig. 6.3 the angle of the principal stress direction α_T is defined as the angle between the x_2-axis and the maximum principal stress T_{22}. Analogically the angle between the vertical axis and the principal strain rate D_1 is defined as α_D.

For the elastoplastic calculation, it can be seen in Fig. 6.5 that the stress and strain rate are not coaxial at the beginning. However, the directions converge with ongoing shearing and are practically the same after a long shearing. The angle α_T is for $K_T < 1$ always smaller than α_D, what is in good agreement with the results of the discrete element computations.

For a simple shear calculation with Hypoplasticity, the coaxiality of stress and deformation rate tensor is also not given (cf. Fig. 6.6) at the beginning. Just as in the elastoplastic calculations after a long shearing the stress tensor and deformation rate tensor are coaxial.

Eq. (6.6) is limited to the case of an earth pressure coefficient at rest $K_0 < 1$, which should be applicable for most slopes. The case $K_0 > 1$ is studied in the following.

Numerical simple shear experiments with discrete elements [116] showed for $K_T = 2$ that the angle of non-coaxiality starts with $\iota > 0$ and then strongly decreases to $\iota = 8°$ at the shear stress peak of the simple shear experiment, which would yield to a slightly higher slope angle due to (6.27), than those

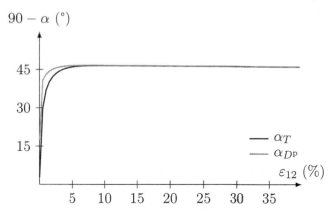

Figure 6.5: Principal directions of the stress tensor α_T and of the strain rate tensor α_{D^P} according to an elastoplastic calculation ($\varphi = 30°, \psi = \varphi/4$)

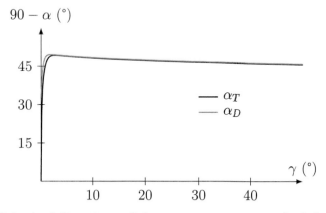

Figure 6.6: Principal directions of the stress tensor α_T and of the deformation tensor α_D according to a hypoplastic calculation ($T_{11} = T_{33} = 4.85\,\text{kPa}, T_{22} = 10\,\text{kPa}, e = 0.667$, Hostun Sand)

calculated with a coaxial flow rule. However, the computed peak shear stress in the simple shear test appears to be approximately the same for $K_T = 2$ and $K_T = 1$, where coaxiality $\iota = 0$ holds approximately throughout the entire shear deformation, which is a consequence of the different angles of dilatancy at the peak ψ in both situations. Hence, the values for ς computed with a coaxial flow rule for $K_T > 1$ seem to be an acceptable approximation.

It is also possible to obtain a higher φ for an overconsolidated soil in elastoplastic calculations. Therefore, it is necessary for the lateral stress at the start to be higher than at the end of a normally consolidated calculation. The lateral stress T_{11} is in this case

$$T_{11} = T_{\mathrm{m}} + T_{\mathrm{q}} \cos 2\alpha_T \quad . \tag{6.28}$$

The relation of T_{11} and T_{22} can be calculated with (6.23) and $\alpha_T = 45° - \psi/2$ to

$$K_T = \frac{T_{11}}{T_{22}} \geq \frac{T_{\mathrm{m}} + T_{\mathrm{q}} \cos 2\alpha_T}{T_{\mathrm{m}} - T_{\mathrm{q}} \cos 2\alpha_T} = \frac{1 + \sin\varphi \sin\psi}{1 - \sin\varphi \sin\psi} \quad . \tag{6.29}$$

If the lateral stress is also higher than in the second solution of (6.22) ($\alpha_T = \varphi/2 - 45°$) the shear strain reaches a maximum for the given friction angle. The lateral stress coefficient has to be higher than

$$K_T = \frac{T_{11}}{T_{22}} \geq \frac{T_{\mathrm{m}} + T_{\mathrm{q}} \cos 2\alpha_T}{T_{\mathrm{m}} - T_{\mathrm{q}} \cos 2\alpha_T} = \frac{1 + \sin^2\varphi}{\cos^2\varphi} \tag{6.30}$$

and the relation between shear and normal stress is $\tan\varphi$, if (6.26) is used, what also means $\phi = \varphi$.

In Figure 6.7, stress-strain-curves for the three different lateral stress coefficient K_T are plotted. It can be seen that for a normal consolidated soil (red line, $K_0 = 1 - \sin\varphi$) the shear stress increases until the maximum is reached. For a coefficient higher than in (6.29) (blue line, $K_T > 1.13$ for this pair of φ and ψ) a peak can be reached, which is higher than in a normal consolidated soil. After the peak, the shear stress decreases again until the value of the normal consolidated soil is reached. In the case of a lateral stress coefficient also higher than (6.30) (in this case $K_T > 1.67$) a peak value of $\tan\varphi$ can be reached (green line). After that, the shear stress also decreases to the normal consolidated value.

Note, that simple shear computations with Hypoplasticity do not confirm an increasing ϕ with increasing K_T, neither for initially loose samples Fig. 6.8 nor for initially dense samples Fig. 6.9. For calculations with Hypoplasticity, the initial K_T-value has no influence on ϕ. Hence, it is not recommended to use the higher ϕ predicted for $K_T > 1$ by calculations with the Mohr-Coulomb elastoplastic model in slope design.

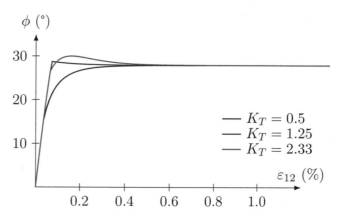

Figure 6.7: Stress-strain curves for different values of K_T ($\varphi = 30°$ and $\psi = \varphi/4$) with an elastoplastic model

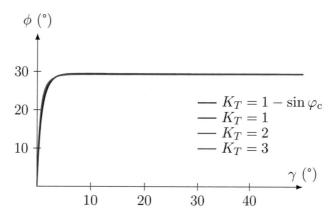

Figure 6.8: Stress-strain curves for different values of K_T and loose Hostun sand ($T_{22} = -200\,\text{kPa}$, $e_0 = 0.822\,\text{kPa}$) with Hypoplasticity

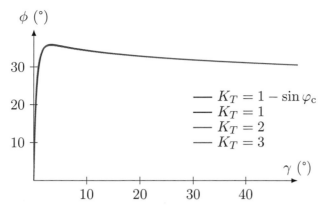

Figure 6.9: Stress-strain curves for different values of K_T and dense Hostun sand ($T_{22} = -100\,\text{kPa}$, $e_0 = 0.627\,\text{kPa}$) with Hypoplasticity

6.3 Limit state function

A limit state function LS depending on the friction angle φ, the dilatancy angle ψ and the inclination of the slope ς can be established such that at an ultimate state, the value of this function is zero. This limit state function reads

$$LS = \tan\varsigma - \frac{\sin\varphi\cos\psi}{1 - \sin\varphi\sin\psi} \tag{6.31}$$

for equation (6.6). For $LS(\varphi, \psi, \varsigma) < 0$ the slope is defined as stable and a state $LS(\varphi, \psi, \varsigma) > 0$ is not feasible, see Fig. 6.10. The line in Fig. 6.10 defines the ultimate state of the slope $LS(\varphi, \psi = 0, \varsigma) = 0$. A point above the line $LS = 0$ results in a stable slope, whereas a point below (shaded area) results in a failure of the slope. In all figures $\varphi = \varsigma$ is plotted as a reference. In Fig. 6.11, the limit state functions for several dilatancy angles are shown.

Equation (6.6) was introduced by Davis [21] to incorporate non-associated plasticity in a slip-line analysis. He proposed to use the reduced-strength parameters

$$\tan\varphi^\star = \frac{\sin\varphi\cos\psi}{1 - \sin\varphi\sin\psi} \quad \text{and} \tag{6.32}$$

$$c^\star = c\frac{\cos\varphi\cos\psi}{1 - \sin\varphi\sin\psi} \tag{6.33}$$

in combination with an associated flow rule. The same relations were proposed by Drescher and Detournay [24] for the use in translational failure mechanisms, when a coaxial flow rule is used. The reduced strength parameters have also been recently introduced in slope stability analysis by means of finite element

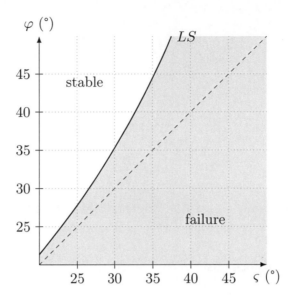

Figure 6.10: Limit state of a cohesionless infinite slope, β is the inclination, φ the friction angle and ψ the dilatancy angle. $LS = 0$ defines the ultimate state (c.f. (6.31)), the filled area is the state of failure.

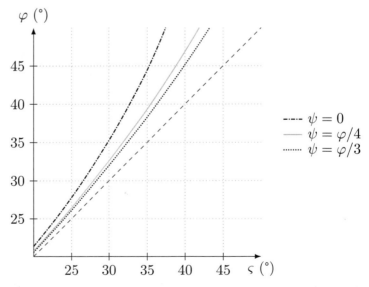

Figure 6.11: Limit states $LS = 0$ of a cohesionless infinite slope in an elasto-plastic material model with Mohr-Coulomb yield function, Eq. (6.31).

limit analysis and finite element strength reduction techniques e.g. Tschuch-nigg *et al.* [121, 122, 123]. The slope angle at limit state is then equal to the reduced friction angle: $\varsigma = \varphi^\star$.

The dilatancy angle ψ is zero for critical states and (6.6) yields $\tan \varsigma = \sin \varphi$. This slope angle is much smaller than the one of the classical approach ($\varsigma = \varphi$). Also for other dilatancy angles ψ smaller than the friction angle φ (6.6) lead to inclinations $\varsigma < \varphi$, cf. Fig 6.11. Only for the cases $\psi = \varphi$, i.e. an associated flow rule, the inclination is at the limit state $\varsigma = \varphi$ (dashed line in Fig. 6.11).

However, in an infinite slope plain strain condition can be assumed. For this boundary condition it is known, that the Mohr-Coulomb failure criterion leads to conservative solutions, because it does not consider the intermediate principal stress.

6.4 Matsuoka-Nakai – elastoplastic

A more appropriate failure criterion for plain-strain conditions in soils is the Matsuoka-Nakai criterion [73], which includes the intermediate principal stress and reads

$$\frac{I_1 I_2}{I_3} = k_{MN} \quad , \tag{6.34}$$

where I_1, I_2 and I_3 are the first, second and third invariants of the stress tensor, respectively, which are

$$I_1 = T_1 + T_2 + T_3 \quad , \tag{6.35}$$
$$I_2 = T_1 T_2 + T_2 T_3 + T_3 T_1 \quad \text{and} \tag{6.36}$$
$$I_3 = T_1 T_2 T_3 \quad , \tag{6.37}$$

and k_{MN} is a material parameter related to the friction angle

$$k_{MN} = \frac{9 - \sin^2 \varphi}{1 - \sin^2 \varphi} \quad . \tag{6.38}$$

An intersection of the failure surface of this criterion with a deviatoric plane is shown in Fig. 6.12. The material strength coincides with Mohr-Coulomb in triaxial conditions, whereas a higher strength can be obtained in plane strain conditions, $\varphi_{\mathrm{mob}} > \varphi$, Fig. 6.13.

The maximum inclination of a cohesionless slope can be computed from the results of a simple shear test simulation for constant normal stress T_{22} and plane strain condition [31]. A maximum shear stress $T_{12,f}$ for the applied

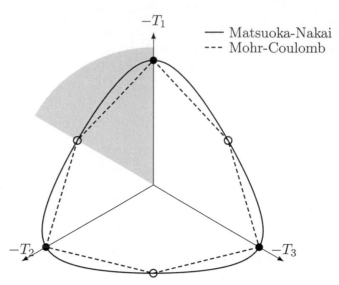

Figure 6.12: Comparison of Matsuoka-Nakai (solid line) and Mohr-Coulomb (dashed line) failure criteria, with marks at triaxial compression (filled circle), triaxial extension (circle) and plane strain states (grey sector)

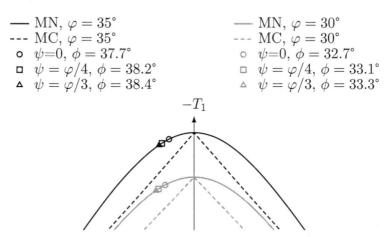

Figure 6.13: End points of simple shear calculations in the principal stress space for different friction angles φ and dilatancy angles ψ.

normal stress can be found from this numerical analysis. It is reasonable to assume that this stress state is equal to the stress state in an infinite slope at the limit state, i.e.

$$\frac{T_{12,f}}{T_{22}} = \tan\varsigma \quad . \tag{6.39}$$

The friction angle φ used in (6.38) can be plotted against ς [105] cf. Fig. 6.14.

Comparing Fig. 6.11 with Fig. 6.14, it can be concluded that in all cases the maximum resulting slope angles are higher when employing the Matsuoka-Nakai failure criterion instead of the Mohr-Coulomb failure criterion. It is possible to achieve an inclination, which is larger than the friction angle of the material, e.g. $\varphi < 25°$ with $\psi = 0$ and $\varphi < 32°$ with $\psi = \varphi/4$

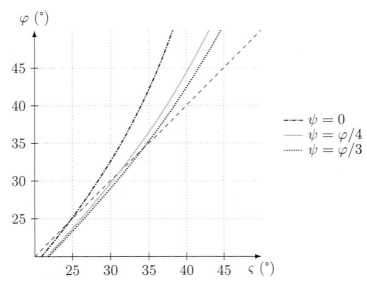

Figure 6.14: Limit state $LS = 0$ for Matsuoka-Nakai failure criterion

The results of the calculation with Matsuoka-Nakai criterion can be approximated with

$$\tan\beta = \frac{\sin(1.085\varphi)\cos\psi}{1 - \sin\varphi\sin\psi} \quad , \tag{6.40}$$

which can be considered as an extension of (6.6). The results of (6.40) agree quite well with the results of the numerical simple shear tests in a range from $\psi = 0$ up to $\psi = \varphi/3$ (max.$\Delta\varphi = \pm 5\%$), Fig. 6.15.

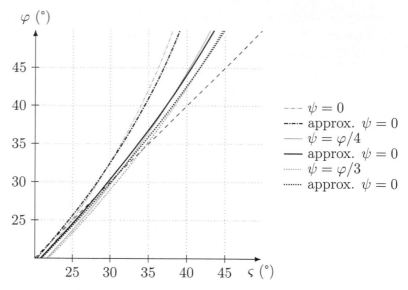

Figure 6.15: Comparison of limit states $LS = 0$ with the approximated relation (6.40) and the elastoplastic model with the Matsuoka-Nakai criterion.

6.5 Matsuoka-Nakai – kinematic

A different approach towards computing the limit state of a slope is to determine the required friction angle for its stability. For this purpose, the full stress tensor has to be calculated in the first step. The shear and normal stresses in every depth of the slope are already known from the classical approach. The remaining stresses need to be determined. Haefeli [50] has proposed a graphical kinematic solution for this task; an analytical method is presented by Fellin [30]. Both methods are based on the assumption that the stress and stretching tensors are coaxial. The principal stresses obtained in [30, 50] are

$$T_1 = \gamma t \cos \varsigma \frac{\cos(\varsigma - \psi) + \sin \varsigma}{\cos \psi} \quad , \tag{6.41}$$

$$T_2 = \gamma t \cos \varsigma \frac{\cos(\varsigma - \psi) + \sin \varsigma \sin \psi}{\cos \psi} \quad \text{and} \tag{6.42}$$

$$T_3 = \gamma t \cos \varsigma \frac{\cos(\varsigma - \psi) - \sin \varsigma}{\cos \psi} \quad . \tag{6.43}$$

Using the principal stresses, the invariants of the stress tensor and k_{MN} of the Matsuoka-Nakai criterion, cf. (6.34), can be calculated,

$$k_{MN} = \frac{(T_1 + T_2 + T_3)(T_1 T_2 + T_2 T_3 + T_3 T_1)}{T_1 T_2 T_3} \quad . \tag{6.44}$$

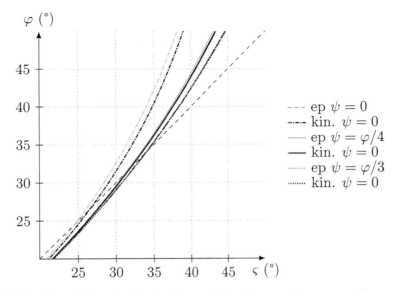

Figure 6.16: Results of the elastoplastic and the kinematic approach

The required friction angle of the material follows from (6.38)

$$\sin \varphi = \sqrt{\frac{k_{MN} - 9}{k_{MN} - 1}} \quad .$$

(6.45)

For the special case of critical state ($\psi = 0$) this can be simplified to [30]

$$\sin \varphi_c = \frac{\sqrt{18 \cos^2 \varsigma - 15 \cos^4 \varsigma - 3}}{5 \cos^2 \varsigma - 1} \quad .$$

(6.46)

The results of the computation of the friction angle required for different dilatancies are shown in Fig. 6.16, which agree quite well with the elastoplastic solution using the Matsuoka-Nakai failure surface.

6.6 Hypoplasticity

Numerical simple shear computations with Hypoplasticity (in particular the version of von Wolffersdorff [126]) can be used to assess the limit state in an infinite slope. Triaxial test calculations with the material parameters summarised in Tab. 6.1 were conducted, e.g. Fig. 6.17. From these tests, the peak friction angle φ_p and the two associated dilatancy measurements ψ and δ are determined.

The friction angle and the dilatancy angle at peak are controlled by the initial isotropic stress $\boldsymbol{T}_{\text{ini}}$ and the initial void ratio e_{ini}. The maximum friction angle

Table 6.1: Material parameters for the hypoplastic constitutive model in version von Wolffersdorff [126] from Herle [51]

Material	φ_c °	h_s GPa	n_H	e_{d0}	e_{c0}	e_{i0}	α_H	β_H
Toyoura sand	30	2.6	0.27	0.61	0.98	1.10	0.18	1.00
Hostun sand	31	1.0	0.29	0.61	0.91	1.09	0.13	2.00
Hochstetten sand	33	1.5	0.28	0.55	0.95	1.05	0.25	1.50
Schlabendorf sand	33	1.6	0.19	0.44	0.85	1.00	0.25	1.00
Hochstetten gravel	36	32.0	0.18	0.26	0.45	0.50	0.10	1.80

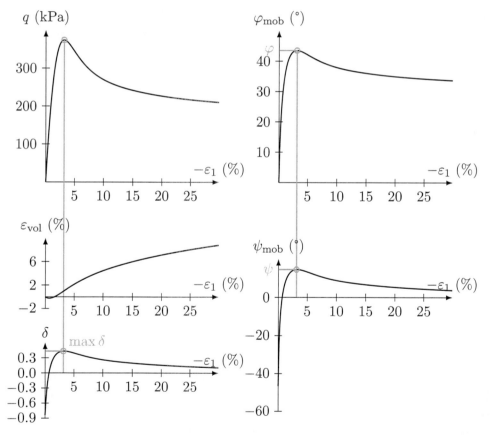

Figure 6.17: Triaxial test with Hostun Sand and initial conditions $e_0 = 0.585$, $T_0 = -84.61\,\mathrm{kPa}$ with Hypoplasticity

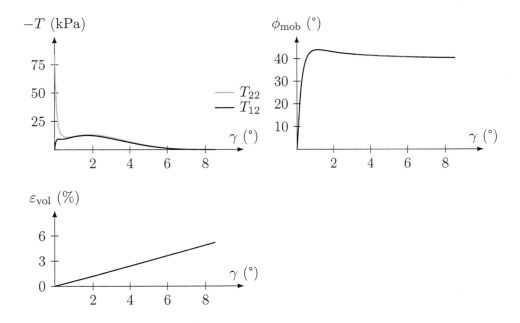

Figure 6.18: Simple shear test with constant δ on Hostun sand and initial conditions $e_0 = 0.585$, $T_{22} = -84.61\,\text{kPa}$ and $T_{11} = T_{22} = -41.03\,\text{kPa}$ (Hypoplasticity)

and the dilatancy at peak increase with a reduction of the stress and reduction in the initial void ratio.

Numerical simple shear test simulations with a constant δ from the peak of the triaxial test and the corresponding initial density e_{ini} are carried out, e.g. Fig. 6.18. For the stress state at the beginning, the same initial vertical stress T_{22} as in the triaxial test calculations is chosen. The lateral stresses (T_{11} and T_{33}) have been determined with Jaky's relation: $(1 - \sin\varphi_c)T_{22}$. The maximum ratio between the shear stress T_{12} and the vertical normal stress T_{22} of this simple shear test can be transformed into a slope inclination ς with the help of (6.39). This ς is plotted in Fig. 6.19, 6.20 and 6.21 together with φ and ψ from the corresponding triaxial calculations for the same material and initial void ratio e.

The solutions of the calculations with Matsuoka-Nakai are added in these figures. The two approaches show similar results. However, the results of the simple shear calculations with Hypoplasticity agree slightly better with those of the kinematic approach in the case of $\psi = 0$. In the calculations with a dilatancy larger than zero, even a larger inclination than with Matsuoka-Nakai is achievable.

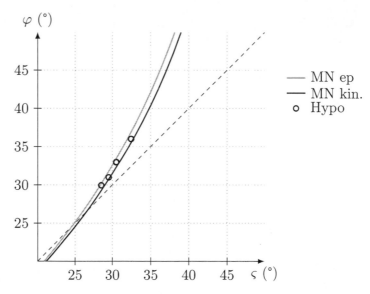

Figure 6.19: Comparison of the different computations for $\psi = 0$

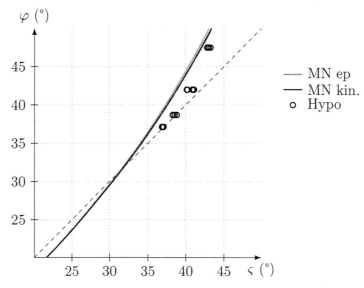

Figure 6.20: Comparison of the different computations for $\psi = \varphi/4$

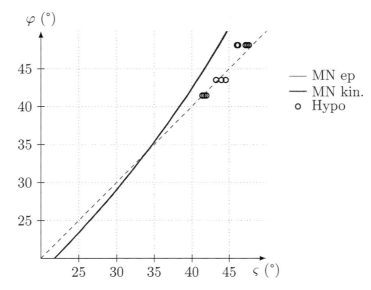

Figure 6.21: Comparison of the different computations for $\psi = \varphi/3$

Table 6.2: Material parameters for Barodesy in the version of Medicus *et al.* [79]

Material	φ_c °	N	λ^*	κ^*	Parameter from
London clay	22.6	1.375	0.11	0.016	Mašín [74]
Weald clay	24	0.8	0.059	0.018	Mašín [72]
Prackovice clay	28.8	1.236	0.101	0.01	Gudehus *et al.* [48]
Hongkong clay	31.4	1.192	0.106	0.01	Gudehus *et al.* [48]
Dresden clay	22.6	0.8	0.059	0.018	Medicus *et al.* [77]

6.7 Barodesy

The maximum inclination of a slope for Barodesy can be calculated in the same manner as for Hypoplasticity. In the following calculations Barodesy in the version of Medicus *et al.* [79] is used, due the fact that only the critical state parameters are needed in this version. The used parameters can be seen in Tab. 6.2.

Triaxial simulations have been carried out with various initial void ratios e_{ini} and different initial stresses $\boldsymbol{T}_{ini}$. One of the results can be seen in Fig. 6.22. From these tests the friction angle in the peak state φ is determined and the corresponding dilatancy measurements ψ and δ. Note, in the case of Barodesy the maximum values of ψ and δ do not occur at the same strain as

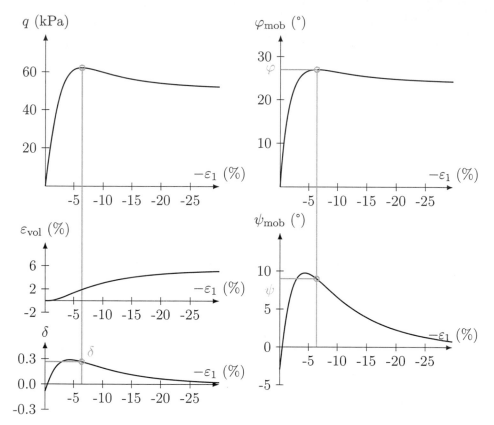

Figure 6.22: Triaxial test with Weald clay and initial conditions $e_0 = 0.6$, $T_0 = -37.46\,\text{kPa}$ (Barodesy)

the maximum friction angle φ.

With the initial void ratio e_{ini} and the vertical stress T_{22} from the triaxial test numerical simple shear with a constant δ from the peak of the triaxial test were conducted. The lateral stresses (T_{11} and T_{33}) have been determined with Jaky's relation. Fig. 6.23 shows results of the simple shear corresponding to the triaxial test in Fig. 6.22.

In these simple shear calculations, the stresses decrease due to the volume increasing strain path. The plateau in the γ-ϕ_{mob}-diagram results from the asymptotic approach of the stress to the proportional stress path. In Fig. 6.24, 6.25 and 6.26 the slope inclination ς from (6.39) is plotted together with φ and ψ from the corresponding triaxial test simulations. The solutions of the calculations with Hypoplasticity and the kinematic approach with the Matsuoka-Nakai failure criteria are added to these figures. The results of Barodesy are in a quite good agreement with the hypoplastic results.

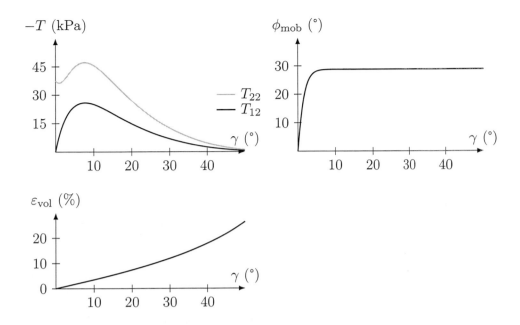

Figure 6.23: Simple shear test with constant δ on Hostun sand and initial conditions $e_0 = 0.6$, $T_{22} = -37.46\,\text{kPa}$ and $T_{11} = T_{22} = -22.22\,\text{kPa}$ (Barodesy)[2]

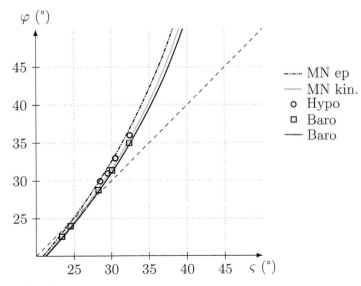

Figure 6.24: Comparison of the different computations for $\psi = 0$

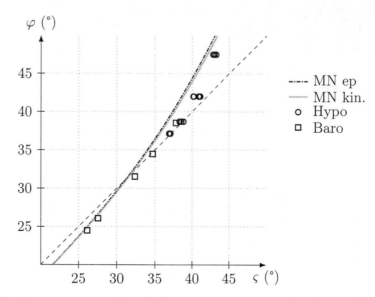

Figure 6.25: Comparison of the different computations for $\psi = \varphi/4$

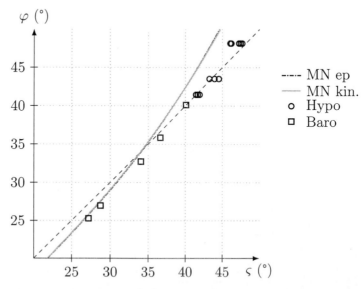

Figure 6.26: Comparison of the different computations for $\psi = \varphi/3$

Because of the asymptotic behaviour, for the Barodesy a simple equation can be derived for the case of $\psi = 0$. In this case, the "peak" friction angle φ_p is the same as the critical friction angle φ_c and the dilatancy δ is zero. The stretching of this simple shear test is then

$$\mathbf{D} = \begin{bmatrix} 0 & D_{12} & 0 \\ D_{12} & 0 & 0 \\ 0 & 0 & 0 \end{bmatrix} \quad . \tag{6.47}$$

This stretching defines an attractor in the stress space, which is the maximum stress ratio ϕ that cannot be exceeded.

To calculate this attractor the fact that $\mathbf{D}$ is not a diagonal matrix has to be considered. For the matrix exponential of a non-diagonal matrix

$$- \exp\left(\alpha \mathbf{D}^0 \right) = -\mathbf{V} \exp\left(\alpha \mathbf{D}^{0\prime} \right) \mathbf{V}^{-1} \tag{6.48}$$

holds true, where $\mathbf{D}^{0\prime}$ is a diagonal matrix with the eigenvalues of $\mathbf{D}^0$ and $\mathbf{V}$ is the matrix assembled from the corresponding eigenvectors. Note, in the case that $\mathbf{V}$ is orthogonal $\mathbf{V}^{-1}$ can be substituted by $\mathbf{V}^{\mathsf{T}}$.

For the tensor in (6.47) this results in

$$\mathbf{D}^{0\prime} = \frac{1}{\sqrt{2}} \begin{bmatrix} -1 & 0 & 0 \\ 0 & 1 & 0 \\ 0 & 0 & 0 \end{bmatrix} \quad \text{and} \quad \mathbf{V} = \frac{1}{\sqrt{2}} \begin{bmatrix} -1 & 1 & 0 \\ 1 & 1 & 0 \\ 0 & 0 & \sqrt{2} \end{bmatrix} \quad .$$

In the case of $\delta = 0$ the value of $\alpha = \alpha_c = \sqrt{2/3} \ln K_c$ in the $\mathbf{R}$-function what leads to

$$\mathbf{R}\left(\mathbf{D}^{0\prime}\right) = - \begin{bmatrix} \exp\left(-\frac{\ln K_c}{\sqrt{3}}\right) & 0 & 0 \\ 0 & \exp\left(\frac{\ln K_c}{\sqrt{3}}\right) & 0 \\ 0 & 0 & 1 \end{bmatrix} = - \begin{bmatrix} K_c^{-\frac{1}{\sqrt{3}}} & 0 & 0 \\ 0 & K_c^{\frac{1}{\sqrt{3}}} & 0 \\ 0 & 0 & 1 \end{bmatrix} \quad . \tag{6.49}$$

After this result is rotated back into the original coordinate system, we get the solution for the tensor in (6.47)

$$\mathbf{R}\left(\mathbf{D}^0\right) = -\frac{1}{2} \begin{bmatrix} K_c^{\frac{1}{\sqrt{3}}} + K_c^{-\frac{1}{\sqrt{3}}} & K_c^{\frac{1}{\sqrt{3}}} - K_c^{-\frac{1}{\sqrt{3}}} & 0 \\ K_c^{\frac{1}{\sqrt{3}}} - K_c^{-\frac{1}{\sqrt{3}}} & K_c^{\frac{1}{\sqrt{3}}} + K_c^{-\frac{1}{\sqrt{3}}} & 0 \\ 0 & 0 & 1 \end{bmatrix} \quad . \tag{6.50}$$

---

[2] The volumetric strain increases because it is plotted over shear angle

From $\boldsymbol{R}$ the friction angle ϕ can be derived as

$$\tan\phi = \left| \frac{R_{12}}{R_{12}} \right| = \left| \frac{K_c^{\frac{1}{\sqrt{3}}} - K_c^{-\frac{1}{\sqrt{3}}}}{K_c^{\frac{1}{\sqrt{3}}} - K_c^{-\frac{1}{\sqrt{3}}}} \right| = \frac{2}{K_c^{\frac{2}{\sqrt{3}}} + 1} - 1 \tag{6.51}$$

with

$$K_c = \frac{1 - \sin\varphi_c}{1 + \sin\varphi_c} \quad . \tag{6.52}$$

This solution is added to Fig. 6.24 and is quite similar to the solution obtained with the kinematic method and the failure surface according to Matsuoka-Nakai (6.34).

6.8   Triaxial test versus simple and direct shear test

The maximum shear stress $T_{12,f}$ measured in a direct shear test is used to determine the friction angle

$$\tan\phi = \frac{T_{12,f}}{T_{22}} \quad . \tag{6.53}$$

However, this value is not generally equal to the friction angle φ determined from a triaxial test, cf. Rowe [98]. In the following plane strain experiments we denote $\varphi_{\mathrm{ps}} := \max\varphi_{\mathrm{mob}}$, which may be higher than φ (cf. Fig. 6.13).

Vardoulakis and Goldscheider [124] concluded from biaxial tests, that the Coulomb friction angle ϕ in the shear band at peak strength is approximately equal to the maximum of the mobilized plane strain Mohr-Coulomb friction angle φ_{ps}, which is larger than the Mohr-Coulomb friction angle φ in triaxial tests due to the plane strain conditions in biaxial tests. The simple shear experiments of Budhu [12] showed that the plane strain Mohr-Coulomb friction angle is higher than the Coulomb friction angle. This is confirmed numerically with discrete elements for simple shear [116] and with Hypoplasticity [62]. Numerical direct shear tests with discrete elements [115] also reveal a plane strain Mohr-Coulomb friction angle higher than the Coulomb friction angle. The mechanics discussed above for the infinite slope can be applied if we assume a simple shear deformation in the shear gap. The relation between these two friction angles is in analogy to (6.6)

$$\tan\phi = \frac{\sin\varphi\cos\psi}{1 - \sin\varphi\sin\psi} \tag{6.54}$$

for a Mohr-Coulomb failure criterion, i.e. $\phi = \varsigma$.

The boundary conditions of a direct shear test are equal to the boundary conditions of an infinite slope. The relation $\phi = \varsigma$ holds irrespective of the material model used for a transformation of ϕ to the material parameter of that model. It therefore seems more appropriate to use direct shear tests to determine a calculation parameter ϕ for a limit state analysis than the triaxial test, since ϕ includes the effect of the dilatancy. This calculated parameter ϕ plays the same role as the reduced friction angle $\varphi^\star$ from (6.32) (Davis [21]). However, the grain size limitation must be considered. The shear gap should be $(10 \ldots 20) \cdot d_{50}$ [108], where d_{50} is the diameter corresponding to 50 % finer in the particle size distribution. Other effects concerning the construction are discussed by Goldscheider [39] and Lindemann [69].

6.9 Impact on limit state analyses

Theoretically, reduced strength parameters (6.32) and (6.33) should be used in any limit state analysis based on slip lines in soil, i.e. also in the common Bishop's method [10] or rigid body calculations [37]. However, based on our investigations we recommend:

1. Whenever possible, the friction angle ϕ derived in a direct shear test in plane strain calculations is to be used

$$\varphi^\star = \phi = \arctan \frac{T_{12,f}}{T_{22}} \quad . \tag{6.55}$$

   It is important to be aware of the limitations of such tests, e.g. the maximum grain size of the soil sample.

2. When a friction angle φ is derived from a triaxial test then

$$\varphi^\star = \min \left\{ \arctan \frac{\sin(1.085\varphi)\cos\psi}{1 - \sin\varphi\sin\psi}, \varphi \right\} \tag{6.56}$$

   can be used if plane strain conditions are assured. That requires the determination of the angle of dilatancy ψ which is not standard in all laboratories. Without knowledge of ψ, the approximation $\varphi^\star \approx \varphi$ may be used for $\varphi < 35°$.

3. If plane strain conditions are not or only partly fulfilled, a conservative estimate can be obtained with

$$\varphi^\star = \arctan \frac{\cos\psi\sin\varphi}{1 - \sin\psi\sin\varphi} \quad , \tag{6.57}$$

   cf. (6.6) and [24, 21].

6.10   Conclusion

The use of reduced shear parameters has been proposed by several authors for limit state analysis based on slip lines. However, the increased soil strength in plane strain condition is not considered by using these parameters, since they are based on a Mohr-Coulomb failure criterion. The here shown calculations suggest, that the plane strain condition somehow counterbalances the effect of the non-associated flow rule, at least for moderate friction angles φ less than $35°$ and moderate stress levels which allow for a dilatancy angle at peak larger than $\varphi/3$. In such cases, the strength parameters determined in a standard triaxial test can be used directly in the common limit state calculation methods of geotechnical engineering without taking the dilatancy into account. This verifies the common practice in geotechnical engineering. In particular, the classical limit state relation for infinite slopes $\varsigma = \varphi$ holds approximately.

Chapter 7

Conclusion

7.1 Summary

The influence of constitutive models on the outcome of geotechnical calculations is not negligible. Hence, the constitutive relation should be chosen with care. The used model has to be able to predict realistic stress-strain-relations for the given task. Even in the case that (some) laboratory tests can be modelled quite satisfactory, this does not imply that also the soil behaviour during geotechnical works with its complex stress paths are predicted correctly, as has been show in chapter 5.

In this thesis, Barodesy for sand has been further developed. Therefore, the R-function, which builds the core of Barodesy, has been replaced by the R-function from the Barodesy version for clay and afterwards it was possible to simplify this function. With the new formulation, which can be used for sand and clay, a large step in the direction of a unified model has been undertaken.

With the new R-function an adjustment of the scalar functions in Barodesy is necessary. All the adjustments in these functions have been made with respect to create a unified model for sand and clay. For the stiffness function h, Ohde's equation is now used, as for clay, and a critical state line is proposed, which is affine to an isotropic compression line. This causes that the function g can be formulated quite similar to the clay version.

It could also be shown that Barodesy, like Hypoplasticity, is able to simulate volumetric deformation under principal stress rotation, for which the invariants of the stress tensor remain constant. Further research is still necessary for Barodesy to predict the correct volumetric deformation. Not every elastoplastic model is able to reproduce such a volumetric change (e.g. an elastoplastic model with a yield surface according to Mohr-Coulomb formulated in the principal stress space or Hardening Soil), however, this defect can be improved as Ishihara and Towhata [55] has shown.

For simpler cases of deformation, like the simple shear test Barodesy shows similar results for limit states as Hypoplasticity and an elastoplastic constitutive relation with a yield criterion according to Matsuoka-Nakai. The results of this test can be compared with the failure of an infinite slope. Due to the fact of its formulation and the asymptotic behaviour, a very simple equation can be derived for the maximum inclination of a slope with a soil in a loose state.

7.2   Limitations

Although Barodesy for sand could be improved, there are still some limitations and some issues, which should be further improved.

- There is no calibration method for some constants in the improved version of Barodesy yet. This shortcoming should be rectified soon in order to apply the constitutive model in practice.

- Reloading and cyclic behaviour is modelled quite poor (ratcheting) like in the other Barodesy versions and Hypoplasticity without intergranular strain.

- For the case that peak states in triaxial test simulations are predicted correctly, the oedometric unloading shows an increase of the lateral stress. This behaviour is unrealistic. If the material parameters are calibrated to predict the stress path in the oedometric unloading correctly, the peak states are underestimated.

- The response envelops show tips in the direction of the isotropic unloading; such a large increase of the stiffness in a very narrow range seems not realistic.

- One shortcoming, which Barodesy shares with many other advanced soil models, is that it still has different formulations for sand and clay. Although the now improved version for sand has already many similarities to the version for clay, further research is necessary for a unified model.

7.3   Outlook

To get rid of the afore mentioned limitations the following steps should be performed:

- Development of a robust and simple calibration method, which uses (in the best case) only standard laboratory tests, like triaxial tests or oedometric tests, such that the material parameters for a soil can easily be calibrated and the model can be used in practical applications.

- To improve the cyclic behaviour (and hence also the reloading) a modification of the basic structure of Barodesy may be necessary. For the reloading a further state variable can be introduced similar the intergranular strain concept. Note, there is already a project funded by the Austrian Science Fund, which tackles this problem for the clay version. The result of this project should be incorporated into the sand or a unified version.

- The influence of the scalar functions f and g on the shape of the response envelops should be investigated in more detail. This may help to eliminate the tips at the response envelops and to correct the unrealistic unloading behaviour in oedometric tests.

- One main goal of further developments should be to unify the different versions for clay and sand. With the here shown improvements a first step has been undertaken, but several further steps are necessary.

- Very important is the availability of a user-defined material subroutine for commercial finite element software like Abaqus and Plaxis. There are already some user-defined material subroutines for various versions of Barodesy. Without such a subroutine, a constitutive relation cannot be used by any practitioner and the entire constitutive model is useless for the community.

Bibliography

[1] A. Al-Tabbaa and D. M. Wood, "An experimentally based" bubble'model for clay," in *Numerical models in geomechanics. NUMOG III*, eds. S. Pietruszczak and G. N. Pande, pp. 91–99, Elsevier Applied Science, London; New York, 1989.

[2] J. Argyris, G. Faust, J. Szimmat, E. Warnke and K. Willam, "Recent developments in the finite element analysis of prestressed concrete reactor vessels," *Nuclear Engineering and Design*, vol. 28, no. 1, pp. 42–75, 1974, doi:10.1016/0029-5493(74)90088-0.

[3] J. R. F. Arthur, K. S. Chua, T. Dunstan and J. I. Rodriguez, "Principal Stress Rotation: A Missing Parameter," *Journal of the Geotechnical Engineering Division, Proceedings of ASCE*, vol. 106, no. 4, pp. 419–433, 1980.

[4] E. Bauer, "Calibration of a Comprehensive Hypoplastic Model for Granular Materials." *Soils and Foundations*, vol. 36, no. 1, pp. 13–26, 1996, doi:10.3208/sandf.36.13.

[5] E. Bauer, "Conditions for embedding Casagrande's critical states into hypoplasticity," *Mechanics of Cohesive-frictional Materials*, vol. 5, no. 2, pp. 125–148, 2000, doi:10.1002/(sici)1099-1484(200002)5:2<125::aid-cfm85>3.0.co;2-0.

[6] K. Been and M. G. Jefferies, "A state parameter for sands," *Géotechnique*, vol. 35, no. 2, pp. 99–112, 1985, doi:10.1680/geot.1985.35.2.99.

[7] T. Benz, *Small-strain stiffness of soils and its numerical consequences*, no. 55 in Mitteilung des Instituts für Geotechnik, IGS, Stuttgart, 2007, zugl.: Stuttgart, Univ., Diss., 2006.

[8] K. Bergholz, "Experimentelle Bestimmung von nichtlinearen Spannungsgrenzbeziehungen," Master's thesis, Technische Universität Dresden, 2009.

[9] K. Bergholz and I. Herle, "Experimentelle Bestimmung der Nichtlinearität von Spannungsgrenzbedingungen im Bereich geringer Spannungen," *geotechnik*, vol. 40, no. 2, pp. 119–125, 2017, doi:10.1002/gete.201600008.

[10] A. Bishop, "The use of the slip circle in the stability analysis of slopes." in *Proceedings of European conference on stability of earth slopes. Stockholm*, pp. 7–17, 1954.

[11] B. Broms and A. Casberian, "Effects of Rotation of the Principal Stress Axes and of the Intermediate Principal Stress on the Shear Strength," in *Proceedings of the Sixth International Conference on Soil Mechanics and Foundation Engineering*, vol. I, pp. 179 – 183, University of Toronto Press, 1965.

[12] M. Budhu, "Lateral Stresses Observed in Two Simple Shear Apparatus," *Journal of Geotechnical Engineering*, vol. 111, no. 6, pp. 698–711, 1985, doi:10.1061/(asce)0733-9410(1985)111:6(698).

[13] R. Butterfield, "A natural compression law for soils (an advance on e-log p/)," *Géotechnique*, vol. 29, no. 4, pp. 469–480, 1979, doi:10.1680/geot.1979.29.4.469.

[14] Y. Canepa and J. Garnier, "Études expérimentales du comportement des fondations superficielles – État de l'art," in *Fondations superficielles - FONDSUP2003*, eds. J.-P. Magnan and N. Droniuc, pp. 155 – 260, Presses de l'ENPC/LCPC, Paris, 2003.

[15] C.-H. Chen, W. Fellin, A. Ostermann and F. Schranz, "Discussion on "Numerical study on finite element implementation of hypoplastic models" by Yutang Ding, Wenxiong Huang, Daichao Sheng, and Scott W. Sloan [Comput. Geotech. 68 (2015) 78–90]," *Computers and Geotechnics*, vol. 71, pp. 276 – 277, 2016, doi:10.1016/j.compgeo.2015.09.005.

[16] J. Chu and S.-C. R. Lo, "Asymptotic behaviour of a granular soil in strain path testing," *Géotechnique*, vol. 44, no. 1, pp. 65–82, 1994, doi: 10.1680/geot.1994.44.1.65.

[17] C. A. Coulomb, "Sur une application des règles de maximis et minimis à quelques problèmes de statique relatifs à l'architecture," *Mémoires de mathématique et de physique, presentés à l'Académie royale des sciences, par divers sçavans & lûs dans ses assemblées*, vol. 7, pp. 343–382, 1773.

[18] R. O. Cudmani, *Statische, alternierende und dynamische Penetration in nichtbindigen Böden*, Veröffentlichungen des Institutes für Bodenmechanik und Felsmechanik am Karlsruher Institut für Technologie (KIT), Karlsruhe: Inst. für Bodenmechanik und Felsmechanik, 2001.

[19] Y. F. Dafalias and M. T. Manzari, "Simple Plasticity Sand Model Accounting for Fabric Change Effects," *Journal of Engineering Mechanics*,

vol. 130, no. 6, pp. 622–634, 2004, doi:10.1061/(ASCE)0733-9399(2004)
130:6(622).

[20] Y. F. Dafalias, A. G. Papadimitriou and X. Li, "Sand Plasticity Model
Accounting for Inherent Fabric Anisotropy," *Journal of Engineering Mechanics*, vol. 130, no. 11, pp. 1319–1333, 2004, doi:10.1061/(ASCE)
0733-9399(2004)130:11(1319).

[21] E. H. Davis, "Theories of plasticity and failure of soil masses," in *Soil
mechanics: selected topics*, ed. I. K. Lee, pp. 341–354, Elsevier, New
York, 1968.

[22] J. Desrues, B. Zweschper and P. Vermeer, "Database for tests on Hostun RF sand," Institutsbericht 13, Institut für Geotechnik, Universität
Stuttgart, 2000.

[23] J. K. Dienes, "On the analysis of rotation and stress rate in deforming
bodies," *Acta Mechanica*, vol. 32, no. 4, pp. 217–232, 1979, doi:10.1007/
bf01379008.

[24] A. Drescher and E. Detournay, "Limit load in translational failure mechanisms for associative and non-associative materials," *Géotechnique*,
vol. 43, no. 3, pp. 443–456, 1993, doi:10.1680/geot.1993.43.3.443.

[25] J. M. Duncan and C.-Y. Chang, "Nonlinear Analysis of Stress and Strain
in Soils," *Journal of the Soil Mechanics and Foundation Division*, vol. 96,
no. 5, pp. 1629–1653, 1970.

[26] F. Engels, *Ludwig Feuerbach und der Ausgang der klassischen deutschen
Philosophie mit Anhang: Karl Marx über Feuerbach vom Jahre 1845*, J.
H. W. Dietz, 1888.

[27] F. Engels and L. Austin, *Feuerbach: the roots of the socialist philosophy*,
Charles H. Kerr & Company, Chicago, 1903.

[28] W. Fellin, *Rütteldruckverdichtung als plastodynamisches Problem (Deep
vibrocompaction as plastodynamic problem)*, Advances in Geotechnical
Engineering and Tunnelling, vol. 2, Balkema, 2000.

[29] W. Fellin, "Ambiguity of safety definition in geotechnical models," in *An-
alyzing Uncertainty in Civil Engineering*, eds. W. Fellin, H. Lessmann,
M. Oberguggenberger and R. Vieider, pp. 71 – 31, Springer-Verlag, 2005,
doi:10.1007/b138177.

[30] W. Fellin, "Abschätzung der Standsicherheit von annähernd unendlich langen Kriechhängen," *geotechnik*, vol. 34, no. 1, pp. 22–31, 2011, doi: 10.1002/gete.201000018.

[31] W. Fellin, "The rediscovery of infinite slope model/Die Wiederentdeckung der unendlich langen Böschung," *Geomechanics and Tunnelling*, vol. 7, no. 4, pp. 299–305, 2014, doi:10.1002/geot.2014000019.

[32] W. Fellin and D. Kolymbas, "Hypoplastizität für leicht Fortgeschrittene," *Bautechnik*, vol. 79, no. 12, pp. 830–841, 2002, doi:10.1002/bate.200205940.

[33] W. Fellin and A. Ostermann, "The critical state behaviour of barodesy compared with the Matsuoka-Nakai failure criterion," *International Journal for Numerical and Analytical Methods in Geomechanics*, vol. 37, no. 3, pp. 299–308, 2011, doi:10.1002/nag.1111.

[34] A. Gajo and D. Muir Wood, "Severn-Trent sand: a kinematic-hardening constitutive model: the q-p formulation," *Géotechnique*, vol. 49, no. 5, pp. 595–614, 1999, doi:10.1680/geot.1999.49.5.595.

[35] German Geotechnical Society, ed., *Empfehlungen des Arbeitskreises Numerik in der Geotechnik - EANG*, Wiley-VCH Verlag GmbH, 2014, doi: 10.1002/9783433604489.

[36] M. Goldscheider, "Grenzbedingung und Fließregel von Sand," *Mechanics Research Communications*, vol. 3, no. 6, pp. 463–468, 1976, doi:10.1016/0093-6413(76)90037-9.

[37] M. Goldscheider, "Standsicherheitsnachweise mit zusammengesetzten Starrkörper-Bruchmechanismen," *geotechnik*, vol. 2, no. 3, pp. 130–139, 1979.

[38] M. Goldscheider, *Spannungen in Sand bei räumlicher, monotoner Verformung*, no. 92 in Veröffentlichungen des Instituts für Bodenmechanik und Felsmechanik der Universität Fridericiana in Karlsruhe, Inst. für Bodenmechanik u. Felsmechanik d. Univ. Fridericiana, 1983.

[39] M. Goldscheider, *Vergleichende Versuche mit einem konventionellen und einem parallel geführten Rahmenschergerät als Grundlage für DIN 18137-3*, Fraunhofer IRB Verlag, Stuttgart, 2003.

[40] M. Goldscheider, "Gültigkeitsgrenzen des statischen Kollapstheorems der Plastomechanik für Reibungsböden," *geotechnik*, vol. 36, no. 4, pp. 243–263, 2013, doi:10.1002/gete.201300007.

[41] A. Grammatikopoulou, "Development, implementation and application of kinematic hardening models for overconsolidated clays," phdthesis, Imperial College London, 2004.

[42] A. Grammatikopoulou, L. Zdravkovic and D. M. Potts, "The effect of the yield and plastic potential deviatoric surfaces on the failure height of an embankment," *Géotechnique*, vol. 57, no. 10, pp. 795–806, 2007, doi:10.1680/geot.2007.57.10.795.

[43] A. E. Green and P. M. Naghdi, "A general theory of an elastic-plastic continuum," *Archive for Rational Mechanics and Analysis*, vol. 18, no. 4, pp. 251–281, 1965, doi:10.1007/bf00251666.

[44] G. Gudehus, "A comparison of some constitutive laws for soils under radially symmetric loading and unloading," in *Proc. Third International Conference on Numerical Methods in Geomechanics*, vol. 4, ed. W. Wittke, pp. 1309–1323, Aachen, 1979.

[45] G. Gudehus, "A comprehensive constitutive equation for granular materials," *Soils Found*, vol. 36, no. 1, pp. 1–12, 1996.

[46] G. Gudehus, *Physical Soil Mechanics*, Springer Berlin Heidelberg, 2011, doi:10.1007/978-3-540-36354-5.

[47] G. Gudehus and D. Mašín, "Graphical representation of constitutive equations," *Géotechnique*, vol. 59, no. 2, pp. 147–151, 2009, doi:10.1680/geot.2007.00155.

[48] G. Gudehus *et al.*, "The soilmodels.info project," *International Journal for Numerical and Analytical Methods in Geomechanics*, vol. 32, no. 12, pp. 1571–1572, 2008, doi:10.1002/nag.675.

[49] M. E. Gurtin and K. Spear, "On the relationship between the logarithmic strain rate and the stretching tensor," *International Journal of Solids and Structures*, vol. 19, no. 5, pp. 437–444, 1983, doi: 10.1016/0020-7683(83)90054-9.

[50] R. Haefeli, "Schneemechanik mit Hinweisen auf die Erdbaumechanik. Beiträge zur Geologie der Schweiz, Geotechnische Serie, Hydrologie, Lieferung 3," , 1939.

[51] I. Herle, *Hypoplastizität und Granulometrie einfacher Korngerüste, Veröffentlichung des Institutes für Bodenmechanik und Felsmechanik*, vol. 142, Karlsruhe: Inst. für Bodenmechanik und Felsmechanik, 1997.

[52] I. Herle, "On selection of advanced constitutive models for geotechnical tasks," in *EUROMECH Colloquium 572 Constitutive Modelling of Soil and Rock*, eds. G. Hofstetter, A. Gajo and D. Kolymbas, Innsbruck, 2016.

[53] D. Hight, T. Henderson, A. Pickles and S. Marchand, "The Nicoll Highway Collapse. Lecture at Meeting of the Hellenic Society of Soil Mechanics and Geotechnical Engineering (HSSMGE)," , 2009.

[54] D. W. Hight, A. Gens and M. J. Symes, "The development of a new hollow cylinder apparatus for investigating the effects of principal stress rotation in soils," *Géotechnique*, vol. 33, no. 4, pp. 355–383, 1983, doi: 10.1680/geot.1983.33.4.355.

[55] K. Ishihara and I. Towhata, "Sand response to cyclic rotation of principal stress directions as induced by wave loads." *Soils Found*, vol. 23, no. 4, pp. 11–26, 1983, doi:10.3208/sandf1972.23.4_11.

[56] J. Jáky, "The Coefficient of Earth Pressure at Rest (Hungarian)," *Magyar Mérnök és Építész Egylet Közlönye*, vol. 78, no. 22, pp. 355 – 358, 1944.

[57] N. Janbu, "Soil Compressibility as Determined by Oedometer and Triaxial Tests," in *Proceedings of the European Conference on Soil Mechanics and Foundation Engineering*, vol. 1, pp. 19–25, Wiesbaden, 1963.

[58] H. Joer, J. Lanier, J. Desrues and E. Flavigny, ""1γ2ε": A New Shear Apparatus to Study the Behavior of Granular Materials," *Geotechnical Testing Journal*, vol. 15, no. 2, pp. 129 – 137, 1992, doi:10.1520/GTJ10235J.

[59] H. A. Joer, J. Lanier and M. Fahey, "Deformation of granular materials due to rotation of principal axes," *Géotechnique*, vol. 48, no. 5, pp. 605–619, 1998, doi:10.1680/geot.1998.48.5.605.

[60] D. Kolymbas, "A rate-dependent constitutive equation for soils," *Mechanics Research Communications*, vol. 4, no. 6, pp. 367–372, 1977.

[61] D. Kolymbas, "A generalized hypoelastic constitutive law," in *Proc. XI Int. Conf. Soil Mechanics and Foundation Engineering, San Francisco*, vol. 5, p. 2626, Balkema, Rotterdam, 1985.

[62] D. Kolymbas, *Eine konstitutive Theorie für Böden und andere körnige Stoffe, Veröffentlichungen des Instituts für Bodenmechanik und*

Felsmechanik der Universität Fridericiana in Karlsruhe, vol. 109, Institut für Bodenmechanik und Felsmechanik der Universität Fridericiana, 1988.

[63] D. Kolymbas, "An outline of hypoplasticity," *Archive of Applied Mechanics*, vol. 61, no. 3, pp. 143–151, 1991, doi:10.1007/BF00788048.

[64] D. Kolymbas, "Introduction to barodesy," *Géotechnique*, vol. 65, no. 1, pp. 52–65, 2015, doi:10.1680/geot.14.p.151.

[65] D. Kolymbas, W. Fellin, B. Schneider-Muntau, G. Medicus and F. Schranz, "Zur Rolle der Materialmodelle beim Standsicherheitsnachweis," *geotechnik*, vol. 39, no. 2, pp. 89–97, 2016, doi:10.1002/gete. 201500017.

[66] D. Kolymbas and G. Medicus, "Genealogy of hypoplasticity and barodesy," *International Journal for Numerical and Analytical Methods in Geomechanics*, vol. 40, no. 18, pp. 2532–2550, 2016, doi:10.1002/nag. 2546.

[67] P. V. Lade, "Effects of Consolidation Stress State on Normally Consolidated Clay," in *Proceedings of NGM-2000 : XIII Nordiska Geoteknikermötet*, ed. H. Rathmayer, Building Information Ltd., Helsinki, 2000.

[68] X. S. Li and Y. Wang, "Linear Representation of Steady-State Line for Sand," *Journal of Geotechnical and Geoenvironmental Engineering*, vol. 124, no. 12, pp. 1215–1217, 1998, doi:10.1061/(ASCE)1090-0241(1998) 124:12(1215).

[69] M. Lindemann, "Vergleichsversuche mit Rahmenschergeräten unterschiedlicher Bauart," *geotechnik*, vol. 26, no. 1, pp. 27–32, 2003.

[70] M. T. Manzari and Y. F. Dafalias, "A critical state two-surface plasticity model for sands," *Géotechnique*, vol. 47, no. 2, pp. 255–272, 1997, doi: 10.1680/geot.1997.47.2.255.

[71] D. Mašín, "Laboratory and numerical modelling of natural clays," Mphil thesis, City University, London, 2004.

[72] D. Mašín, "Clay hypoplasticity with explicitly defined asymptotic states," *Acta Geotechnica*, vol. 8, no. 5, pp. 481–496, 2013, doi:10.1007/ s11440-012-0199-y.

[73] H. Matsuoka and T. Nakai, "Stress-deformation and strength characteristics of soil under three different principal stresses," *Proceedings of the*

Japan Society of Civil Engineers, vol. 1974, no. 232, pp. 59–70, 1974, doi:10.2208/jscej1969.1974.232_59.

[74] D. Mašín, "A hypoplastic constitutive model for clays," *International Journal for Numerical and Analytical Methods in Geomechanics*, vol. 29, no. 4, pp. 311–336, 2005, doi:10.1002/nag.416.

[75] G. Medicus, *Barodesy and its Application for Clay*, no. 20 in Advances in Geotechnical Engineering and Tunneling, Logos, Berlin, 2015.

[76] G. Medicus and W. Fellin, "An improved version of barodesy for clay," *Acta Geotechnica*, vol. 12, no. 2, pp. 365–376, 2016, doi:10.1007/s11440-016-0458-4.

[77] G. Medicus, W. Fellin and D. Kolymbas, "Barodesy for clay," *Géotechnique Letters*, vol. 2, no. 4, pp. 173–180, 2012, doi:10.1680/geolett.12.00037.

[78] G. Medicus, W. Fellin and F. Schranz, "Konzepte der Barodesie," *Bautechnik*, vol. 95, no. 9, pp. 620 – 638, 2018, doi:10.1002/bate.201800015.

[79] G. Medicus, D. Kolymbas and W. Fellin, "Proportional stress and strain paths in barodesy," *International Journal for Numerical and Analytical Methods in Geomechanics*, vol. 40, no. 4, pp. 509–522, 2015, doi:10.1002/nag.2413.

[80] M. Mergili and W. Fellin, "Slope stability and geographic information systems: an advanced model versus the infinite slope stability approach," in *Problems of Decrease in Natural Hazards and Risks, The International Scientifically-Practical Conference GEORISK*, pp. 119–124, 2009.

[81] M. Mergili, W. Fellin, S. M. Moreiras and J. Stötter, "Simulation of debris flows in the Central Andes based on Open Source GIS: possibilities, limitations, and parameter sensitivity," *Natural Hazards*, vol. 61, no. 3, pp. 1051–1081, 2012, doi:10.1007/s11069-011-9965-7.

[82] M. Mergili, I. Marchesini, M. Rossi, F. Guzzetti and W. Fellin, "Spatially distributed three-dimensional slope stability modelling in a raster GIS," *Geomorphology*, vol. 206, pp. 178–195, 2014, doi:10.1016/j.geomorph.2013.10.008.

[83] K. Miura, S. Miura and S. Toki, "Deformation behavior of anisotropic dense sand under principal stress axes rotation." *Soils and Foundations*, vol. 26, no. 1, pp. 36–52, 1986, doi:10.3208/sandf1972.26.36.

[84] D. Muir Wood, *Soil Behaviour and Critical State Soil Mechanics*, Cambridge University Press, Cambridge [u.a.], 1990.

[85] D. Muir Wood, *Geotechnical modelling*, New York, NY Spon Press, 2004.

[86] P. J. Naughton and B. C. O'Kelly, "Stress non-uniformity in a hollow cylinder torsional sand specimen," *Geomechanics and Geoengineering*, vol. 2, no. 2, pp. 117–122, 2007, doi:10.1080/17486020701377124.

[87] L. Naujoks, H. Muhs and J. Siegfried, *Flachgründungen: Grundbruch und Setzungen*, Berichte aus der Bauforschung, Wilhelm Ernst und Sohn, 1963.

[88] A. Niemunis and I. Herle, "Hypoplastic model for cohesionless soils with elastic strain range," *Mechanics of Cohesive-frictional Materials*, vol. 2, no. 4, pp. 279–299, 1997, doi:10.1002/(SICI)1099-1484(199710)2: 4<279::AID-CFM29>3.0.CO;2-8.

[89] R. Nova, "Controllability of the incremental Response of Soil Specimen subjected toArbitrary Loading Programmes," *Journal of the Mechanical Behavior of Materials*, vol. 5, no. 2, 1994, doi:10.1515/JMBM.1994.5.2. 193.

[90] J. Ohde, "Zur Theorie der Druckverteilung im Baugrund," *Der Bauingenieur*, vol. 20, pp. 451–459, 1939.

[91] R. H. G. Parry, "Triaxial Compression and Extension Tests on Remoulded Saturated Clay," *Géotechnique*, vol. 10, no. 4, pp. 166–180, 1960, doi:10.1680/geot.1960.10.4.166.

[92] J. M. Pestana and A. J. Whittle, "Compression model for cohesionless soils," *Géotechnique*, vol. 45, no. 4, pp. 611–631, 1995, doi:10.1680/geot. 1995.45.4.611.

[93] L. Prandtl, "Über die Härte plastischer Körper," *Nachrichten von der Königlichen Gesellschaft der Wissenschaften zu Göttingen, Mathematisch-Physikalische Klasse*, pp. 74–85, 1920.

[94] L. Rendulic, "Ein Grundgesetz der Tonmechanik und sein experimenteller Beweis," *Der Bauingenieur*, vol. 18, no. 31|32, pp. 459–467, 1937.

[95] O. Reynolds, "LVII. On the dilatancy of media composed of rigid particles in contact. With experimental illustrations," *The London, Edinburgh, and Dublin Philosophical Magazine and Journal of Science*, vol. 20, no. 127, pp. 469–481, 1885, doi:10.1080/14786448508627791.

[96] K. H. Roscoe, A. N. Schofield and C. P. Wroth, "On The Yielding of Soils," *Géotechnique*, vol. 8, no. 1, pp. 22–53, 1958, doi:10.1680/geot. 1958.8.1.22.

[97] P. Rowe, "The Stress-Dilatancy Relation for Static Equilibrium of an Assembly of Particles in Contact," *Proceedings of the Royal Society A: Mathematical, Physical and Engineering Sciences*, vol. 269, no. 1339, pp. 500–527, 1962, doi:10.1098/rspa.1962.0193.

[98] P. Rowe, "The Relation Between the Shear Strength of Sands in Triaxial Compression, Plane Strain and Direct Shear," *Géotechnique*, vol. 19, no. 1, pp. 75–86, 1969.

[99] A. Sayao and Y. Vaid, "A critical assessment of stress nonuniformities in hollow cylinder test specimens." *Soils and Foundations*, vol. 31, no. 1, pp. 60–72, 1991, doi:10.3208/sandf1972.31.60.

[100] T. Schanz, *Zur Modellierung des mechanischen Verhaltens von Reibungsmaterialien*, no. 45 in Mitteilungen des Instituts für Geotechnik, Stuttgart, Institut für Geotechnik, Stuttgart, Stuttgart, 1998, zugl.: Stuttgart, Univ., Habilschr., 1998.

[101] T. Schanz, P. Vermeer and P. Bonnier, "The hardening soil model: Formulation and verification," in *Beyond 2000 in computational geotechnics: 10 years of PLAXIS International ; proceedings of the International Symposium Beyond 2000 in Computational Geotechnics, Amsterdam, the Netherlands, 18 - 20 March 1999*, ed. R. B. J. Brinkgreve, pp. 281–296, Balkema, Rotterdam, 1999.

[102] G. Schneebeli, "Une analogie mécanique pour les terres sans cohésion," *Comptes rendus hebdomadaires des séances de l'Académie des sciences*, vol. 243, pp. 125 – 126, 1956.

[103] B. Schneider-Muntau, G. Medicus and W. Fellin, "Strength reduction method in Barodesy," *Computers and Geotechnics*, vol. 95, pp. 57–67, 2018, doi:10.1016/j.compgeo.2017.09.010.

[104] B. Schneider-Muntau, F. Schranz and W. Fellin, "The possibility of a statistical determination of characteristic shear parameters from triaxial tests," *Beton- und Stahlbetonbau*, vol. 113, pp. 86–90, 2018, doi:10.1002/best.201800038.

[105] F. Schranz, "Standsicherheit von unendlich langen Hängen," Master's thesis, Universität Innsbruck, 2014.

[106] F. Schranz and W. Fellin, "Stability of infinite slopes investigated with elastoplasticity and hypoplasticity," *geotechnik*, vol. 39, no. 3, pp. 184–194, 2016, doi:10.1002/gete.201500021.

[107] E. Schwiteilo and I. Herle, "Vergleichsstudie zur Kompressibilität und zu den Scherparametern von Ton aus Ödometer- und Rahmenscherversuchen," *geotechnik*, vol. 40, no. 3, pp. 204–217, 2017, doi:10.1002/gete.201600015.

[108] S. Shibuya, T. Mitachi and S. Tamate, "Interpretation of direct shear box testing of sands as quasi-simple shear," *Géotechnique*, vol. 47, no. 4, pp. 769–790, 1997, doi:10.1680/geot.1997.47.4.769.

[109] S. Sloan, "Geotechnical stability analysis," *Géotechnique*, vol. 63, no. 7, pp. 531–571, 2013, doi:10.1680/geot.12.RL.001.

[110] M. Taiebat and Y. F. Dafalias, "SANISAND: Simple anisotropic sand plasticity model," *International Journal for Numerical and Analytical Methods in Geomechanics*, vol. 32, no. 8, pp. 915–948, 2008, doi:10.1002/nag.651.

[111] F. Tatsuoka, T. B. S. Pradhan and N. Horii, "Discussion on the paper by Jewell, R. A. and Wroth, C. P. (37-1, pp. 53-68)," *Géptechnique*, vol. 38, no. 1, pp. 148–153, 1988, doi:10.1680/geot.1988.38.1.139.

[112] D. D. Taylor, *Fundamentals of Soil Mechanics*, New York: John Wiley & Sons, London: Chapman & Hall, 1948.

[113] K. Terzaghi and R. B. Peck, *Die Bodenmechanik in der Baupraxis*, Springer-Verlag, 1961.

[114] J. A. M. Teunissen and S. E. J. Spierenburg, "Stability of infinite slopes," *Géotechnique*, vol. 45, no. 2, pp. 321–323, 1995, doi:10.1680/geot.1995.45.2.321.

[115] C. Thornton and L. Zhang, "Numerical Simulations of the Direct Shear Test," *Chemical Engineering & Technology*, vol. 26, no. 2, pp. 153–156, 2003, doi:10.1002/ceat.200390022.

[116] C. Thornton and L. Zhang, "A numerical examination of shear banding and simple shear non-coaxial flow rules," *Philosophical Magazine*, vol. 86, no. 21-22, pp. 3425–3452, 2006, doi:10.1080/14786430500197868.

[117] Z.-X. Tong, J.-M. Zhang, Y.-L. Yu and G. Zhang, "Drained Deformation Behavior of Anisotropic Sands during Cyclic Rotation of Principal Stress

Axes," *Journal of Geotechnical and Geoenvironmental Engineering*, vol. 136, no. 11, pp. 1509–1518, 2010, doi:10.1061/(ASCE)GT.1943-5606. 0000378.

[118] M. Topolnicki, *Observed stress-strain behaviour of remoulded saturated clay and examination of two constitutive models*, no. 107 in Veröffentlichungen des Instituts für Bodenmechanik und Felsmechanik der Universität Fridericiana in Karlsruhe, Inst. für Bodenmechanik u. Felsmechanik d. Univ. Fridericiana, Karlsruhe, 1987.

[119] M. Topolnicki, G. Gudehus and B. Mazurkiewicz, "Observed stress-strain behaviour of remoulded saturated clay under plane strain conditions," *Géotechnique*, vol. 40, no. 2, pp. 155–187, 1990, doi:10.1680/ geot.1990.40.2.155.

[120] C. Truesdell and W. Noll, *The Non-Linear Field Theories of Mechanics*, Springer Berlin Heidelberg, 2nd edn., 1992, doi:10.1007/ 978-3-662-13183-1.

[121] F. Tschuchnigg, H. Schweiger and S. Sloan, "Slope stability analysis by means of finite element limit analysis and finite element strength reduction techniques. Part I: Numerical studies considering non-associated plasticity," *Computers and Geotechnics*, vol. 70, pp. 169–177, 2015, doi: 10.1016/j.compgeo.2015.06.018.

[122] F. Tschuchnigg, H. Schweiger and S. Sloan, "Slope stability analysis by means of finite element limit analysis and finite element strength reduction techniques. Part II: Back analyses of a case history," *Computers and Geotechnics*, vol. 70, pp. 178–189, 2015, doi:10.1016/j.compgeo.2015.07. 019.

[123] F. Tschuchnigg, H. Schweiger, S. Sloan, A. Lyamin and I. Raissakis, "Comparison of finite-element limit analysis and strength reduction techniques," *Géotechnique*, vol. 65, no. 4, pp. 249–257, 2015, doi: 10.1680/geot.14.p.022.

[124] I. Vardoulakis and M. Goldscheider, "Biaxialgerät zur Untersuchung der Festigkeit und Dilatanz von Scherfugen in Böden," *geotechnik*, vol. 3, no. 1, pp. 19–31, 1980.

[125] R. Verdugo and K. Ishihara, "The Steady State of Sandy Soils." *Soils and Foundation*, vol. 36, no. 2, pp. 81–91, 1996, doi:10.3208/sandf.36.2_81.

[126] P.-A. von Wolffersdorff, "A hypoplastic relation for granular materials with a predefined limit state surface," *Mechanics of Cohesive-frictional Materials*, vol. 1, no. 3, pp. 251–271, 1996, doi:10.1002/(sici)1099-1484(199607)1:3<251::aid-cfm13>3.0.co;2-3.

[127] A. J. Whittle and R. V. Davies, "Nicoll Highway Collapse: Evaluation of Geotechnical Factors Affecting Design of Excavation Support System," in *Proceedings of the International Conference on Deep Excavations*, 2006.

[128] D. Wijewickreme, "Behaviour of sand under simultaneous increase in stress ratio and rotation of principal stresses," phdthesis, University of British Columbia, 1990, doi:10.14288/1.0050457.

[129] Z. X. Yang, X. S. Li and J. Yang, "Undrained anisotropy and rotational shear in granular soil," *Géotechnique*, vol. 57, no. 4, pp. 371–384, 2007, doi:10.1680/geot.2007.57.4.371.

[130] H.-S. Yu, *Plasticity and Geotechnics, Advances in Mechanics and Mathematics*, vol. 13, Springer, New York, 2006.

[131] H.-S. Yu and X. Yuan, "On a class of non-coaxial plasticity models for granular soils," *Proceedings of the Royal Society A: Mathematical, Physical and Engineering Sciences*, vol. 462, no. 2067, pp. 725–748, 2006, doi:10.1098/rspa.2005.1590.

[132] L. Zhao, F. Yang, Y. Zhang, H. Dan and W. Liu, "Effects of shear strength reduction strategies on safety factor of homogeneous slope based on a general nonlinear failure criterion," *Computers and Geotechnics*, vol. 63, pp. 215–228, 2015, doi:10.1016/j.compgeo.2014.08.015.

Appendix A

Glossary

a	Vector with the deformation functions, Sec. 5.4
$\dot{a}$	Vector with the derivatives of the deformation, Sec. 5.4
B	Right Cauchy-Green deformation tensor, Sec. 5.4
b	scalar quantity in the barodesy, Sec. 3.3
b_T	intermediate principal stress parameter $b_T = \frac{T_2 - T_3}{T_1 - T_3}$, Sec. 5.6
C^e	Elastic stiffness matrix, Sec. 6.2
C^{ep}	Elastoplastic stiffness matrix, Sec. 6.2
c	Cohesion of the soil, Sec. 2.0
c_e	ratio between the void ratio at normal compression e_i and the critical void ratio e_c for the same stress level $c_e = \frac{1+e_i}{1+e_c}$, Sec. 4.2
c_u	Undrained cohesion of the soil, Sec. 2.3
c_α	relation between stress ratio in the critical states for compression and extension, Sec. 5.2
D	Rate of deformation tensor, symmetric part of the velocity gradient, Sec. 1.2
D^0	direction of the rate of deformation tensor, Sec. 3.2
D^e	elastic part of the stretching in elasto-plastic constitutive models, Sec. 1.2
D^p	plastic part of the stretching in elasto-plastic constitutive models, Sec. 1.2
D^*	deviatoric part of the stretching, Sec. 5.2
D^{*e}	deviatoric part of the elastic stretching, Sec. 5.2
D_{vol}	volumetric part of the stretching, Sec. 5.2
D_{vol}^e	volumetric part of the elastic stretching, Sec. 5.2
D_{vol}^p	volumetric part of the plastic stretching, Sec. C.3
E	Internal variables, like void ratio, strain history,..., Sec. 5.4
E	Young's module, Sec. 6.2
E_{50}	loading stiffness in triaxial tests for Hardening Soil, Sec. 5.2

E_{ur} unloading and reloading stiffness in triaxial tests for Hardening Soil, Sec. 5.2

e void ratio, ratio between volume of voids V_p and volume of soil grains V_s $e = \frac{V_\text{p}}{V_\text{s}}$, Sec. 3.2

e_c void ratio in the critical state for the actual stress level, Sec. 3.2

e_{c0} critical void ratio at mean stress level of zero, Sec. 3.2

e_i maximum void ratio for isotropic compression at the actual stress level, Sec. 4.2

e_{i0} loosest void ratio for isotropic compression at mean stress level of zero, Sec. 4.2

$\boldsymbol{F}$ Deformation gradient, Sec. 5.3

f scalar function in the Barodesy, Sec. 3.2

f_y Yield function for elastoplastic constitutive models, Sec. 2.3

f_y^s Yield function for shear hardening in Hardening Soil, Sec. 5.2

f_y^c Yield function for compression hardening in Hardening Soil, Sec. 5.2

f^b bounding surface in Sanisand, Sec. 5.2

f^c critical surface in Sanisand, Sec. 5.2

f^d dilatancy surface in Sanisand, Sec. 5.2

G Shear stiffness in (linear) elastic models, Sec. 5.2

g scalar function in the Barodesy, Sec. 3.2

g_y Plastic potential in elastoplastic constitutive models, Sec. 2.3

h scalar function in the Barodesy, Sec. 3.2

h_s Granular stiffness used in Hypoplasticity, Sec. 3.2

K Bulk stiffness, Sec. 3.2

K_0 K_0 loading path from oedometric deformations, earth pressure coefficient ar rest, Sec. 3.1

K_c K_c principal stress ratio in the critical state, it is also the active earth pressure coefficient, Sec. 4.1

K_r reference stiffness, Sec. 3.2

K_T relation between horizontal and vertical normal stress, usually the the relation between maximum and minimal principal stress, Sec. 4.1

$\boldsymbol{L}$ Velocity gradient, Sec. 5.3

LS Limit state function, Sec. 6.3

M slope of the critical state line in the p-q-plane, Sec. 3.2

m	Normalvector on the plastic potential at the actual stress point, Sec. 6.2
m_B	Function in the Barodesy for clay, Sec. 3.3
m_r	Factor to increase the reversal stiffness of Hypoplasticity in the intergranular strain concept, Sec. 5.2
m_S	Describes the opening angle of the cone of the yield surface in Sanisand, a typical value is 5 % of α^c, Sec. 5.2
m_t	Factor to increase the tangential deformation stiffness of Hypoplasticity in the intergranular strain concept, Sec. 5.2
N	specific volume $(1 + e)$ of the isotropic normal compression line at the reference pressure σ^*, Sec. 2.0
n	Normalvector on the yield surface at the actual stress point, Sec. 6.2
n_H	Exponent in the isotropic compression line according to Bauer [4], Sec. 3.2
n_S	Exponent in the yield function of Sanisand, Sec. 5.2
p	mean effective stress $p = \frac{T_1 + T_2 + T_3}{3}$, Sec. 2.3
$\mathring{p}$	Rate of mean stress p, Sec. 5.2
p_0	Isotropic hardening variable in Saninsad, Sec. 5.2
p_{at}	atmospheric pressure as reference pressure, Sec. 3.2
p_r	reference pressure, Sec. 3.2
Q	Rotation matrix of the deformation gradient, Sec. 5.4
q	deviatoric stress $q = \sqrt{\frac{(T_1 - T_2)^2 + (T_2 - T_3)^2 + (T_3 - T_1)^2}{2}}$, for axisymmetric loading $q = T_3 - T_1$, Sec. 2.3
q_a	deviatroic asymptotic stress in Hardening Soil, Sec. 5.2
R	stress rate corresponding to a proportional strain path, Sec. 3.2
R^0	direction of a stress rate corresponding to a proportional strain path, Sec. 3.1
R_v^0	direction vector of a stress rate corresponding to a proportional strain path, Sec. 3.2
R_N	Residual in Newton's method, Sec. 5.4
R	Maximum value of the intergranular strain, Sec. 5.2
r	normalized deviatoric stress in sanisand, Sec. 5.2
T	effective Cauchy stress tensor; compression negative, Sec. 1.2
T_v	Vector pointing from the origin of the stress space to the stress tensor T, Sec. 3.1
T^0	Normalised stress tensor T, Sec. 3.1

T_v^0 direction of the stress tensor T in the principal stress space, Sec. 3.1

T^* Deviatoric part of the stress, Sec. 5.2

$\overset{\circ}{T}$ objective stress rate given by the constitutive model, Sec. 3.0

$\overset{\circ}{T}^0$ direction of the objective stress rate $\overset{\circ}{T}$, Sec. 3.2

$\overset{\circ}{T}_v^0$ direction vector of the objective stress rate $\overset{\circ}{T}$, Sec. 3.2

$\overset{\circ}{T}^*$ Rate of the deviatoric stress, Sec. 5.2

$\dot{T}$ stress rate, Sec. 3.2

$\hat{T}$ Stress tensor normalised by its trace, Sec. C.2

$\hat{T}^*$ Deviatoric part of the normalised stress tensor, Sec. C.2

T_m Mean value of the maximum and minimum principal stress, Sec. 6.2

T_q Half of the difference of the maximum and minimum principal stress, Sec. 6.2

U Right stretch tensor of the deformation gradient, Sec. 5.4

V Matrix of eigenvectors, Sec. 5.4

W Spin tensor, antimetric part of the velocity gradient, Sec. 5.3

X Initial position of the material point, Sec. 5.3

x Actual position of the material point, Sec. 5.3

$\boldsymbol{\alpha}$ Deviatoric back-stress ratio, a hardening paramter in Sanisand, Sec. 5.2

α scalar function in the R-function, Sec. 3.2

α^c Distance of the critical surface in the deviatoric plane and the current loading direction, Sec. A.0

α_c^c value of the critical state surface for axisymmetric compression in Sanisand, Sec. 5.2

α_D Principal direction of the actual stretching D, Sec. 6.2

α_{min} minimal value of α in the R-function, Sec. 4.1

α_T Principal stress direction for the actual stress state T, Sec. 5.5

$\tan\beta$ dilatancy measurement for axisymmetric loading with $\tan\beta = \frac{-\operatorname{tr} D}{D_1}$, Sec. 1.2

γ_{12} Shear strain in the x_1-x_2-plane, $\gamma_{12} = 2\varepsilon_{12}$, Sec. 5.5

$\boldsymbol{\delta}$ Intergranular strain in Hypoplasticity, Sec. 5.2

δ dilatancy measurement based on the normalised D, $\delta = \operatorname{tr} D^0 = D_1^0 + D2^0 + D_3^0$, Sec. 1.2

ε strain tensor; elongation positive, Sec. 1.2

$\dot{\varepsilon}$ strain rate, for rectlinear deformations and logarithmic strain it is the same as $\boldsymbol{D}$, Sec. 1.2

ε^* deviatoric strain tensor, Sec. 5.2

ε_q deviatoric strain, Sec. 3.2

ε_{vol} volumetric strain, Sec. 3.2

ζ Exponent in Barodesy for weightening the pyknotropy, Sec. 4.2

η stress ratio $\eta = \frac{q}{p}$, Sec. 3.2

θ Lode angle, Sec. 5.2

ι Angle of non-coaxiality, i.e. the deviation fo the principal direction of stress α_T and stretching α_D, Sec. 6.2

κ factor between loading and unloading stiffness, Sec. 4.2

κ^* slope of the isotropic unloading line in the p-e-plane, Sec. 3.3

λ_e slope of the critical state line, Sec. 3.2

λ^* slope of the isotropic normal compression line in the p-e-plane, Sec. 3.2

ν Poisson's ratio, Sec. 6.2

ξ Exponent in the stiffness relation of Ohde [90] and Janbu [57], Sec. 3.2

ξ_e Exponent in the critical state line of Li and Wang [68], Sec. 3.2

σ reference pressure for the normal compression line, Sec. 3.2

ς Inclination of a slope, Sec. 2.3

σ^* reference pressure for the normal compression line, Sec. 3.2

φ Mohr-Coulomb friction angle, Sec. 2.0

φ_c critical friction angle after long shearing, Sec. 2.0

φ_{max} theoretical maximum mobilised friction angle $\varphi_{max} = 90°$, Sec. 3.2

φ_{mob} mobilised friction angle $\varphi_{mob} = \frac{T_3 - T_1}{T_1 + T_3}$, Sec. 2.3

φ_p friction angle for peak states, Sec. 3.2

φ_{ps} Mohr-Coulomb friction angle in plane strain condition, Sec. 6.8

ϕ friction angle from simple or direct shear tests, Sec. 6.1

ϕ_{c} critical friction angle derived from a simple or direct shear test after long shearing, Sec. 6.1

ϕ_{mob} mobilised friction angle derived from a simple or direct shear test $\phi_{\mathrm{mob}} = \frac{T_{12}}{T_{11}}$, Sec. 6.1

ϕ_{p} friction angle for peak states in simple or direct shear tests, Sec. 6.1

ψ dilatancy angle derived from Mohr-Coulomb, Sec. 1.2

ψ_D angle in the Rendulic plane for deformations, $\psi_D = 0$ for the isotropic compression axis, Sec. 1.2

ψ_e state paramter in Sanisand, which describes the difference between the actual void ratio and the critical one, Sec. 5.2

ψ_{mob} dilatancy angle derived from Mohr-Coulomb $\sin \psi_{\mathrm{mob}} = \frac{\dot{\varepsilon}_{\mathrm{v}}^{\mathrm{p}}}{-2\dot{\varepsilon}_1^{\mathrm{p}} + \dot{\varepsilon}_{\mathrm{v}}^{\mathrm{p}}}$, where $\dot{\varepsilon}_1^{\mathrm{p}}$ is the loading direction, Sec. 6.1

ψ_{p} dilatancy angle derived from Mohr-Coulomb in the peak state, Sec. 6.1

ψ_T angle in the Rendulic plane for stresses, $\psi_T = 0$ for the isotropic compression axis, Sec. 1.2

$\boldsymbol{\Omega}$ Tensor of the angular velocity $\boldsymbol{\omega} = \dot{\boldsymbol{Q}}\boldsymbol{Q}^{\mathsf{T}}$, Sec. 5.4

Appendix B

Recalculation of a triaxial test

In Figure B.1, triaxial tests of normally consolidated Weald Clay are simulated with the elasto-plastic constitutive model and a yield surface according to Mohr-Coulomb. The model is calibrated according to the consolidated drained triaxial test in Figure B.1a. The parameters are estimated with $\varphi = 23.5°$, $c = 0\,\text{kPa}$ and $\psi = 0°$. Young's module E and Poison's ration ν for the elastic behaviour are $E = 5000\,\text{kPa}$ and $\nu = 0.31$, respectively. With these parameters, which are calibrated on a normal consolidated soil, a cautious estimation has been made. The maximum admissible deviatoric shear stress q_{max} is realistically estimated.

In Figure B.1b, the result of a consolidated undrained triaxial test with a normal-consolidated clay is compared with a elasto-plastic simulation. The maximum deviatoric stress that occurs at the simulation is highly overestimated with the Mohr-Coulomb yield surface despite the conservative choice of parameters. For $\psi > 0°$ the deviatoric stress q would even increase further.

In a conventional consolidated undrained triaxial test the stress and strain states are axial symmetric, i.e. $T_2 = T_3$ and $\varepsilon_2 = \varepsilon_3$. In the following, only triaxial compressions will be considered.

Elastic range In the elastic range the equations of the linear elasticity theory are used. In the case of triaxial tests the stress invariants $p = \dfrac{1}{3}(T_1 + 2T_3)$ and $q = T_3 - T_1$ can be used. These stress rates can be calculated with the elastic stiffness matrix $\boldsymbol{C}^{\text{e}}$

$$\begin{pmatrix} \dot{p} \\ \dot{q} \end{pmatrix} = \boldsymbol{C}^{\text{e}} \begin{pmatrix} \dot{\varepsilon}_{\text{vol}} \\ \dot{\varepsilon}_q \end{pmatrix} \quad \text{where} \quad \boldsymbol{C}^{\text{e}} = \begin{pmatrix} K & 0 \\ 0 & 3G \end{pmatrix} \tag{B.1}$$

The bulk modulus K and the shear modulus G can be calculated with Young's module E and Poison's ratio ν with

$$K = \frac{E}{3(1 - 2\nu)} \quad \text{and} \quad G = \frac{E}{2(1 + \nu)} \quad . \tag{B.2}$$

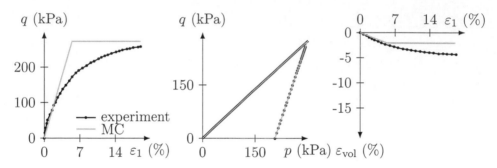

(a) Results of a consolidated drained triaxial test and a simulation with the elasto-plastic model and a yield surface according to Mohr-Coulomb, the maximum deviatoric stresses q_{max} agrees well.

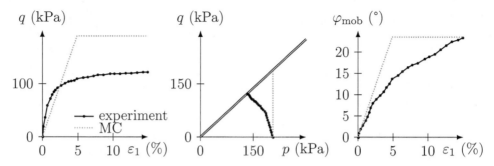

(b) Results of a consolidated undrained triaxial test and a simulation with the elasto-plastic model and a yield surface according to Mohr-Coulomb, the maximum deviatoric stresses q_{max} is overestimated. However, the same mobilised friction angle φ_{mob} or q/p reached.

Figure B.1: Simulations of normal consolidated triaxial tests on Weald Clay (from Mašín [72], data from Parry [91])

This formulas are correct if $f_y < 0$, with

$$f_y(\boldsymbol{T}) = f_y(p, q) = q - Mp \quad \text{with} \quad M = \frac{6 \sin \varphi}{3 - \sin \varphi} \tag{B.3}$$

Plastic range In the case of plastic flow (i.e. $f_y = 0$), the elastoplastic stiffness matrix is defined as follows, see e.g. Muir Wood [85]:

$$\begin{pmatrix} \dot{p} \\ \dot{q} \end{pmatrix} = \boldsymbol{C}^{ep} \begin{pmatrix} \dot{\varepsilon}_{vol} \\ \dot{\varepsilon}_q \end{pmatrix} \quad \text{with} \quad \boldsymbol{C}^{ep} = \boldsymbol{C}^{e} - \frac{\boldsymbol{C}^{e} \dfrac{\partial g_y}{\partial \boldsymbol{T}} \dfrac{\partial f_y}{\partial \boldsymbol{T}}^{\mathsf{T}} \boldsymbol{C}^{e}}{\dfrac{\partial f_y}{\partial \boldsymbol{T}}^{\mathsf{T}} \boldsymbol{C}^{e} \dfrac{\partial g_y}{\partial \boldsymbol{T}}} \tag{B.4}$$

and g_y the plastic potential

$$g_y(\boldsymbol{T}) = g_y(p, q) = q - M^* p \quad \text{with} \quad M^* = \frac{6 \sin \psi}{3 - \sin \psi} \qquad \text{(B.5)}$$

If the stress derivatives of $f_y(\boldsymbol{T})$ and $g_y(\boldsymbol{T})$ are inserted in the elastoplastic stiffness matrix $\boldsymbol{C}^{\text{ep}}$ from (B.4) it can be written as

$$\boldsymbol{C}^{\text{ep}} = \begin{pmatrix} K & 0 \\ 0 & 3G \end{pmatrix} - \frac{1}{KMM^* + 3G} \begin{pmatrix} MM^* K^2 & -3M^* GK \\ -3MGK & 9G^2 \end{pmatrix} \qquad \text{(B.6)}$$

or

$$\boldsymbol{C}^{\text{ep}} = \frac{3GK}{KMM^* + 3G} \begin{pmatrix} 1 & M^* \\ M & MM^* \end{pmatrix} \qquad \text{(B.7)}$$

In the elastic region, the isochoric deformation ($\text{tr}\,\boldsymbol{D} = 0$) leads to $\dot{p} = 0\,\text{kPa}$, cf. (B.1), i.e. the mean pressure remains constant at $p = 206.2\,\text{kPa}$. The deviatoric stress q increases until the stress path reaches the yield condition, cf. (B.1): $\dot{\varepsilon}_q > 0 \rightsquigarrow \dot{q} > 0$.

As soon as plastic flow occurs, the stress rates $\dot{p}$ and $\dot{q}$ are calculated with (B.4) and (B.7). For the conservative estimations with $\psi = 0°$ ($\rightsquigarrow M^* = 0$), the second column of the elastoplsatic stiffness matrix (B.7) consists of zeros. Thus, with $\text{tr}\,\boldsymbol{D} = 0$, both, the change of the mean pressure $\dot{p}$ and the change of the deviatoric stress $\dot{q}$ vanish. The stresses p and q remain constant. For example, the maximal deviatoric stress can be calculated to

$$q_{\max} = Mp = \frac{6 \sin 23.5°}{3 - \sin 23.5°} \cdot 206.2\,\text{kPa} \approx 190\,\text{kPa} \quad ,$$

compare Fig. B.1a.

Isochoric deformation and plane strain

In the following, plane strain conditions are examined for the elastoplastic model with a yield surface according to Mohr-Coulomb. In plane strain $D_2 = 0$, furthermore rectilinear extension will be considered. In the case of isochoric deformation $\dot{\varepsilon}_{\text{vol}} = 0$ and hence $D_1 = -D_3$.

Elastic range The change of stress is calculated with linear elasticity theory

$$\begin{pmatrix} \dot{T}_1 \\ \dot{T}_2 \\ \dot{T}_3 \end{pmatrix} = \boldsymbol{C}^{\text{e}} \begin{pmatrix} D_1 \\ 0 \\ D_3 \end{pmatrix} \quad \text{with} \quad \boldsymbol{C}^{\text{e}} = \frac{E}{(1+\nu)(1-2\nu)} \begin{pmatrix} 1-\nu & \nu & \nu \\ \nu & 1-\nu & \nu \\ \nu & \nu & 1-\nu \end{pmatrix}$$

$$\text{(B.8)}$$

From $\operatorname{tr} \boldsymbol{D} = 0$ (i.e. $D_1 = -D_3$) it follows that the mean pressure is constant

$$\dot{p} = -\frac{\operatorname{tr} \boldsymbol{T}}{3} = \frac{E}{3(1 - 2\nu)}(D_1 + D_3) = 0 \tag{B.9}$$

Plastic range The stress rate is

$$\begin{pmatrix} \dot{T}_1 \\ \dot{T}_2 \\ \dot{T}_3 \end{pmatrix} = \boldsymbol{C}^{\mathrm{ep}} \begin{pmatrix} D_1 \\ 0 \\ D_3 \end{pmatrix} \quad \text{with} \quad \boldsymbol{C}^{\mathrm{ep}} = \boldsymbol{C}^{\mathrm{e}} - \frac{\boldsymbol{C}^{\mathrm{e}} \dfrac{\partial g_{\mathrm{y}}}{\partial \boldsymbol{T}} \dfrac{\partial f_{\mathrm{y}}}{\partial \boldsymbol{T}}^{\mathsf{T}} \boldsymbol{C}^{\mathrm{e}}}{\dfrac{\partial f_{\mathrm{y}}}{\partial \boldsymbol{T}}^{\mathsf{T}} \boldsymbol{C}^{\mathrm{e}} \dfrac{\partial g_{\mathrm{y}}}{\partial \boldsymbol{T}}} \quad . \tag{B.10}$$

Here the yield function $f_{\mathrm{y}}(\boldsymbol{T})$ is

$$f_{\mathrm{y}}(\boldsymbol{T}) = (T_1 + T_3)\sin\varphi - (T_1 - T_3) - 2c\cos\varphi \tag{B.11}$$

and the plastic potential $g_{\mathrm{y}}(\boldsymbol{T})$ is

$$g_{\mathrm{y}}(\boldsymbol{T}) = (T_1 + T_3)\sin\psi - (T_1 - T_3) \quad . \tag{B.12}$$

The elastoplastic stiffness matrix (B.10) can be written for $\psi = 0°$ with the derivatives of f_{y} and g_{y} as

$$\boldsymbol{C}^{\mathrm{ep}} = \frac{E}{2(1+\nu)(1-2\nu)} \begin{pmatrix} 1 - \sin\varphi & 2\nu(1 - \sin\varphi) & 1 - \sin\varphi \\ 2\nu & 2(1 - \nu) & 2\nu \\ 1 + \sin\varphi & 2\nu(1 + \sin\varphi) & 1 + \sin\varphi \end{pmatrix} \tag{B.13}$$

For the case of isochoric deformation (i.e. $\operatorname{tr} \boldsymbol{D} = 0$ and $D_1 = -D_3$) the mean pressure p and the deviatoric stress q remain constant, i.e. $\dot{p} = 0$ and $\dot{q} = 0$ for plastic flow

$$\dot{p} = -\frac{\operatorname{tr} \boldsymbol{T}}{3} = \frac{E}{3(1 - 2\nu)}(D_1 + D_3) = 0 \quad \text{and} \tag{B.14}$$

$$\begin{aligned} \dot{q} &= \sqrt{\frac{1}{2}\left[(\dot{T}_1 - \dot{T}_2)^2 + (\dot{T}_2 - \dot{T}_3)^2 + (\dot{T}_3 - \dot{T}_1)^2\right]} \\ &= \frac{E\sqrt{(1 - 2\nu)^2 + 3\sin^2\varphi}}{(1 + \nu)(1 - 2\nu)}(D_1 + D_3) = 0 \quad . \end{aligned} \tag{B.15}$$

Appendix C

Constitutive Models

C.1 Barodesy

C.1.1 Version of Medicus *et al.* for clay

$$\mathring{\boldsymbol{T}} = h(f\boldsymbol{R}^0 + g\boldsymbol{T}^0)\|\boldsymbol{D}\|$$

$$\boldsymbol{R} = -\exp\left(\alpha\boldsymbol{D}^0\right)$$

$$\alpha(\delta) = \frac{\sqrt{2}\ln K_\delta}{\sqrt{3-\delta^2}}$$

$$K_\delta = 1 - \frac{1}{1 + c_1(m_{\mathrm{B}} - c_2)^2} \quad \text{with} \quad m_{\mathrm{B}} = -\frac{3\delta}{6 - 2\delta^2}$$

$$\delta = \operatorname{tr}\boldsymbol{D}^0$$

$$h = c_3\|\boldsymbol{T}\|^{c_4}$$

$$f = c_6 b\delta - 0.5$$

$$g = (1 - c_6)b\delta + \left(\frac{1+e}{1+e_{\mathrm{c}}}\right)^{c_5} - 0.5$$

$$b = -\frac{1}{c_3\Lambda} + \frac{1}{\sqrt{3}}2^{c_5\lambda^*} - \frac{1}{\sqrt{3}}$$

$$\Lambda = \frac{\kappa^* - \lambda^*}{2\sqrt{3}}\delta + \frac{\lambda^* + \kappa^*}{2}$$

$$e_{\mathrm{c}} = \exp\left(N - \lambda^*\ln\frac{2p}{\sigma^*}\right) - 1$$

Table C.1: Parameter for Weald clay for Barodesy with $\sigma^* = 1\,\text{kPa}$

φ_{c}	N	λ^*	κ^*
24°	0.8	0.059	0.018

Parameter

Constants

$$c_1 = \frac{1 - \sin\varphi_{\text{c}}}{2c_2^2 \sin\varphi_{\text{c}}}$$

$$c_2 = -\frac{3\sqrt{2} + 3}{2} \approx -3.6213$$

$$c_3 = \frac{\frac{\sqrt{3}}{\kappa^*} - \frac{\sqrt{3}}{\lambda^*}}{2^{c_5\lambda^*} + 0.002^{c_5\lambda^*} - 2}$$

$$c_4 = 1$$

$$c_5 = \frac{1 + \sin\varphi_{\text{c}}}{1 - \sin\varphi_{\text{c}}}$$

$$c_6 = \frac{1}{2\left(2^{c_5\lambda^*} - \frac{-\sqrt{3}}{c_3\kappa^*} - 1\right)}$$

C.1.2　Version of Kolymbas for sand

$$\overset{\circ}{\boldsymbol{T}} = h(f\boldsymbol{R}^0 + g\boldsymbol{T}^0)\|\boldsymbol{D}\|$$

$$\boldsymbol{R} = -\exp\left(\alpha\boldsymbol{D}^0\right)$$

$$\alpha(\delta) = c_1 \exp\left(c_2\delta\right)$$

$$\delta = \operatorname{tr}\boldsymbol{D}^0$$

$$h = -\frac{c_4 + c_5\|\boldsymbol{T}\|}{e - e_{\min}}$$

$$f = \delta + c_3 e_{\text{c}}$$

$$g = -c_3 e$$

$$e_{\text{c}} = \frac{e_{\min} + B}{1 - B}$$

$$B = \frac{e_{c0} - e_{min}}{e_{c0} + 1} \left(\frac{c_4 + c_5 \|\boldsymbol{T}\|}{c_4} \right)^{-\frac{1+e_{min}}{c_5}}$$

Parameter

Table C.2: Parameter for Hostun sand for Barodesy

φ_c	c_2	c_3	c_4	c_5	e_{c0}	e_{min}
33.8°	1	2.5076	1 kPa	40	0.91	0.5

Constants

$$c_1 = \sqrt{\frac{2}{3}} \ln \left(\frac{1 - \sin \varphi_c}{1 + \sin \varphi_c} \right)$$

C.1.3 Improved version for sand

$$\mathring{\boldsymbol{T}} = h(f\boldsymbol{R}^0 + g\boldsymbol{T}^0)\|\boldsymbol{D}\|$$

$$\boldsymbol{R} = \exp\left(\alpha \boldsymbol{D}^0\right)$$

$$\alpha(\delta) = -30 + c_3 \frac{|\delta - \sqrt{2}|^{c_2}}{\left(1 + |\delta - \sqrt{2}|\right)^{c_1}}$$

$$\delta = \operatorname{tr} \boldsymbol{D}^0$$

$$h = c_4 \|\boldsymbol{T}\|^{\xi}$$

$$f = c_8 b\delta + c_6$$

$$g = (1 - c_8)b\delta + c_5 \left[\left(\frac{1+e}{1+e_c} \right)^{\zeta} - 1 \right] - c_6$$

$$b = (b_{ext} - b_{comp}) \left(\frac{\delta + \sqrt{3}}{2\sqrt{3}} \right)^{c_7} + b_{comp}$$

$$e_c = (1 + e_{c0}) \exp \left(-\frac{p^{1-\xi}}{c_4(1 - \xi)} \right) - 1$$

Parameter

Table C.3: Parameter for Hostun sand for Barodesy with $p_\mathrm{r} = 1\,\mathrm{kPa}$

φ_c	K_r	ξ	κ	e_{c0}	c_e	c_5	c_6	c_7
33.8°	430 kPa	0.74	3	0.904	1.093	3	5	10

Constants

$$b_\mathrm{comp} = \frac{c_5(c_e{}^\varsigma - 1) - 3}{\sqrt{3}}$$

$$b_\mathrm{ext} = \frac{c_5(1 - c_e{}^\varsigma) - 3\kappa}{\sqrt{3}}$$

$$c_1 = -14.9940 \ln\left(\frac{\alpha_\mathrm{p} + 30}{\alpha_\mathrm{c} + 30}\right) - 14.6975 \ln\left(\frac{\alpha_0 + 30}{\alpha_\mathrm{c} + 30}\right)$$

$$c_2 = \frac{\ln\left(\frac{\alpha_0 + 30}{\alpha_\mathrm{c} + 30}\right) + 0.3466 c_1}{0.5348}$$

$$c_3 = (\alpha_\mathrm{c} + 30)\frac{(1 + \sqrt{2})^{c_1}}{\sqrt{2}^{c_2}}$$

$$c_4 = K_\mathrm{r}\left(\sqrt{3}p_\mathrm{r}\right)^{-\xi}$$

$$c_8 = -\frac{c_6}{\sqrt{3}b_\mathrm{ext}}$$

$$\varsigma = \frac{1}{K_\mathrm{c}} = \frac{1 + \sin\varphi_\mathrm{c}}{1 - \sin\varphi_\mathrm{c}}$$

$$\alpha_\mathrm{p} = \ln\left(\frac{9 - 11\sin\varphi_\mathrm{c}}{9 + 13\sin\varphi_\mathrm{c}}\right)$$

$$\alpha_\mathrm{c} = \frac{\sqrt{2}\ln\left(\frac{1 - \sin\varphi_\mathrm{c}}{1 + \sin\varphi_\mathrm{c}}\right)}{\sqrt{3}}$$

$$\alpha_0 = \ln\left(1 - \sin\varphi_\mathrm{c}\right)$$

C.2 Hypoplasticity

C.2.1 Version of von Wolffersdorff for sand

$$\overset{\circ}{\boldsymbol{T}} = f_{\mathrm{s}}\left(\mathcal{L} : \boldsymbol{D} + \boldsymbol{N}\|\boldsymbol{D}\|\right)$$

$$\mathcal{L} = \frac{1}{\hat{\boldsymbol{T}} : \hat{\boldsymbol{T}}}\left(F^2\mathcal{I} + a_{\mathrm{H}}^2\hat{\boldsymbol{T}} \otimes \hat{\boldsymbol{T}}\right)$$

$$\boldsymbol{N} = \frac{f_{\mathrm{d}}Fa_{\mathrm{H}}}{\hat{\boldsymbol{T}} : \hat{\boldsymbol{T}}}\left(\hat{\boldsymbol{T}} + \hat{\boldsymbol{T}}^*\right)$$

$$a_{\mathrm{H}} = \frac{\sqrt{3}(3 - \sin\varphi_{\mathrm{c}})}{2\sqrt{2}\sin\varphi_{\mathrm{c}}}$$

$$\sqrt{\frac{\tan^2\psi_{\mathrm{H}}}{8} + \frac{2 - \tan^2\psi_{\mathrm{H}}}{2 + \sqrt{2}\tan\psi_{\mathrm{H}}\cos 3\theta}} - \frac{\tan\psi_{\mathrm{H}}}{2\sqrt{2}}$$

$$\tan\psi_{\mathrm{H}} = \sqrt{3}\|\hat{\boldsymbol{T}}^*\|$$

$$f_{\mathrm{s}} = \left(\frac{e_{\mathrm{i}0}e_{\mathrm{c}}}{e_{\mathrm{c}0}e}\right)^{\beta_{\mathrm{H}}}\frac{h_{\mathrm{s}}}{n_{\mathrm{H}}}\frac{1 + e_{\mathrm{i}}}{e_{\mathrm{i}}}\left(\frac{3p}{h_{\mathrm{s}}}\right)^{1 - n_{\mathrm{H}}}\left[3 + a_{\mathrm{H}}^2 - a_{\mathrm{H}}\sqrt{3}\left(\frac{e_{\mathrm{i}0} - e_{\mathrm{d}0}}{e_{\mathrm{c}0} - e_{\mathrm{d}0}}\right)\right]^{-1}$$

$$f_{\mathrm{d}} = \left(\frac{e - e_{\mathrm{d}}}{e_{\mathrm{c}} - e_{\mathrm{d}}}\right)^{\alpha_{\mathrm{H}}}$$

$$\frac{e_{\mathrm{i}}}{e_{\mathrm{i}0}} = \frac{e_{\mathrm{c}}}{e_{\mathrm{c}0}} = \frac{e_{\mathrm{d}}}{e_{\mathrm{d}0}} = \exp\left[-\left(\frac{3p}{h_{\mathrm{s}}}\right)^{n_{\mathrm{H}}}\right]$$

Parameter

Table C.4: Parameter of Hypoplasticity for Hostun sand [51]

φ_{c}	h_{s}	n_{H}	$e_{\mathrm{d}0}$	$e_{\mathrm{c}0}$	$e_{\mathrm{i}0}$	α_{H}	β_{H}
30°	2.6 GPa	0.27	0.61	0.98	1.10	0.18	1.00

C.2.2   Intergranular strain concept

$$\mathring{\boldsymbol{T}} = \mathcal{M} : \boldsymbol{D}$$

$$\mathcal{M} = \left[\rho^\chi m_{\mathrm{t}} + (1 - \rho^\chi) m_{\mathrm{r}}\right] \mathcal{L} +$$
$$\begin{cases} \rho^\chi (1 - m_{\mathrm{t}})\mathcal{L} : \hat{\boldsymbol{\delta}}\hat{\boldsymbol{\delta}} + \rho^\chi \boldsymbol{N}\hat{\boldsymbol{\delta}} & \text{for: } \hat{\boldsymbol{\delta}} : \boldsymbol{D} > 0, \\ \rho\chi(m_{\mathrm{r}} - m_{\mathrm{t}})\mathcal{L} : \hat{\boldsymbol{\delta}}\hat{\boldsymbol{\delta}} & \text{for: } \hat{\boldsymbol{\delta}} : \boldsymbol{D} \le 0, \end{cases}$$

$$\rho = \frac{\|\boldsymbol{\delta}\|}{R}$$

$$\hat{\boldsymbol{\delta}} = \begin{cases} \boldsymbol{\delta}/\|\boldsymbol{\delta}\| & \text{for } \boldsymbol{\delta} \neq \boldsymbol{0} \\ \boldsymbol{0} & \text{for } \boldsymbol{\delta} = \boldsymbol{0} \end{cases}$$

$$\mathring{\boldsymbol{\delta}} = \begin{cases} (\mathcal{I} - \hat{\boldsymbol{\delta}}\hat{\boldsymbol{\delta}}\rho^{\beta_r}) : \boldsymbol{D} & \text{for } \hat{\boldsymbol{\delta}} : \boldsymbol{D} > 0 \\ \boldsymbol{D} & \text{for } \hat{\boldsymbol{\delta}} : \boldsymbol{D} \le 0 \end{cases}$$

C.3   Sanisand

Elastic relations

$$D_{\mathrm{vol}}^{\mathrm{e}} = \frac{\mathring{p}}{K}$$

$$\boldsymbol{D}^{*\mathrm{e}} = \frac{\mathring{\boldsymbol{T}}^*}{2G}$$

$$K = K_{\mathrm{r}} p_{\mathrm{at}} \frac{1 + e}{e} \left(\frac{p}{p_{\mathrm{at}}}\right)^{2/3}$$

$$G = G_0 p_{\mathrm{at}} \frac{(2.97 - e)^2}{1 + e} \sqrt{\frac{p}{p_{\mathrm{at}}}}$$

$$f_{\mathrm{y}} = \frac{3}{2}(\boldsymbol{T}^* - p\boldsymbol{\alpha}) : (\boldsymbol{T}^* - p\boldsymbol{\alpha}) - m_{\mathrm{S}}^2 p^2 \left[1 - \left(\frac{p}{p_0}\right)^{n_{\mathrm{S}}}\right]$$

Plastic relation

$$D_{\text{vol}}^{\text{p}} = \langle L \rangle \left[Dr_{\text{ef}} + \exp(-Vr_{\text{ef}}) \right]$$

$$\boldsymbol{D}^{*\text{P}} = \langle L \rangle \left[\sqrt{\frac{3}{2}} \boldsymbol{n}_{\text{T}} r_{\text{ef}} + \frac{3}{2} X \boldsymbol{r} \exp(-Vr_{\text{ef}}) \right]$$

$$r_{\text{ef}} = \sqrt{\frac{3}{2} (\boldsymbol{r} - \boldsymbol{\alpha}) : (\boldsymbol{r} - \boldsymbol{\alpha})}$$

$$\boldsymbol{r} = \frac{\boldsymbol{T}^*}{p}$$

$$L = \frac{1}{K_{\text{p}}} \left(\frac{\partial f_{\text{y}}}{\partial \boldsymbol{T}} : \mathring{\boldsymbol{T}} \right) = \frac{1}{K_{\text{p}}} \left(\frac{\partial f_{\text{y}}}{\partial \boldsymbol{T}^*} : \mathring{\boldsymbol{T}}^* + \frac{\partial f_{\text{y}}}{\partial p} : \mathring{p} \right)$$

$$K_{\text{p}} = - \left(\frac{\partial f_{\text{y}}}{\partial \boldsymbol{\alpha}} : \bar{\boldsymbol{\alpha}} + \frac{f_{\text{y}}}{\partial p_0} \bar{p}_0 \right)$$

$$\bar{\boldsymbol{\alpha}} = hr_{\text{ef}}(\boldsymbol{\alpha}^{\text{b}} - \boldsymbol{\alpha})$$

$$h_{\text{h}} = \frac{b_0}{\frac{3}{2} \left[(\boldsymbol{b}_{\text{ref}} - (\boldsymbol{\alpha}^{\text{b}} - \boldsymbol{\alpha})) : \boldsymbol{n}_{\text{T}} \right]^2}$$

$$b_0 = G_0 h_0 (1 - c_h e) \sqrt{\frac{p_{\text{at}}}{p}}$$

$$\boldsymbol{b}_{\text{ref}} = \sqrt{\frac{2}{3}} \alpha_{\text{c}}^{\text{b}} (1 + c_\alpha) \boldsymbol{n}_{\text{T}}$$

$$\alpha_{\text{c}}^{\text{b}} = \alpha_{\text{c}}^{\text{c}} \exp(-n^{\text{b}} \psi_e)$$

$$\alpha_{\text{c}}^{\text{d}} = \alpha_{\text{c}}^{\text{c}} \exp(n^{\text{d}} \psi_e)$$

$$\boldsymbol{\alpha}^{\text{c,b,d}} = \sqrt{\frac{2}{3}} \alpha^{\text{c,b,d}} \boldsymbol{n}_{\text{T}}$$

$$\psi_e = e - e_{\text{c}}$$

$$e_{\text{c}} = e_{\text{c0}} - \lambda_e \left(\frac{p}{p_{\text{at}}} \right)^\xi$$

$$\alpha^{\text{c,b,d}} = g_{\text{s}}(\theta, c_\alpha) \alpha_{\text{c}}^{\text{c,b,d}}$$

$$g_{\text{s}}(\theta, c_\alpha) = \frac{2 c_\alpha}{(1 + c_\alpha) - (1 - c_\alpha) \cos 3\theta}$$

$$\boldsymbol{n}_{\text{T}} = \frac{\boldsymbol{r} - \boldsymbol{\alpha}}{\sqrt{(\boldsymbol{r} - \boldsymbol{\alpha}) : (\boldsymbol{r} - \boldsymbol{\alpha})}} = \frac{\boldsymbol{T}^* - p\boldsymbol{\alpha}}{\sqrt{(\boldsymbol{T}^* - p\boldsymbol{\alpha}) : (\boldsymbol{T}^* - p\boldsymbol{\alpha})}}$$

$$\bar{p}_0 = \frac{(1+e)p_0 \exp(-Vr_{\mathrm{ef}})}{e^{\left[\frac{\rho_c - \left(\frac{p_0}{p_{\mathrm{at}}}\right)^{1/3}}{K_{\mathrm{r}}}\right]}(1 - \mathrm{sgn}\,\delta_{\mathrm{s}}|\delta_{\mathrm{s}}|^{\theta_{\mathrm{L}}})}$$

$$\delta_{\mathrm{s}} = 1 - \frac{p}{p_b}\left(1 + 3\frac{\boldsymbol{\alpha}:\boldsymbol{\alpha}}{(g_{\mathrm{s}}\alpha_{\mathrm{c}}^{\mathrm{c}})^2}\right)$$

$$p_b = p_r\left(\frac{1}{e}\right)^{1/\rho_c}$$

$$D = A_d\left(\alpha_d - \sqrt{\frac{3}{2}}\boldsymbol{\alpha}:\boldsymbol{n}\right)$$

Evolution of internal variables

$$\dot{\boldsymbol{\alpha}} = \langle L\rangle\bar{\boldsymbol{\alpha}}$$

$$\dot{p}_0 = \langle L\rangle\bar{p}_0$$

Parameter

Table C.5: Material parameters of Sanisand for Toyura Sand

G_0	K_{r}	$\alpha_{\mathrm{c}}^{\mathrm{c}}$	c_α	$e_{\mathrm{c}0}$	λ_e	ξ_e	n^{d}
125 kPa	150 kPa	1.2	0.712	0.934	0.019	0.7	2.1

A_{d}	n^{b}	h_0	c_{h}	p_r	ρ_{c}	θ_{L}	X
0.4	1.25	36.96	0.987	5.5 MPa	0.37	0.18	0.8

Constants

Table C.6: Constants of Sanisand as defined in Taiebat and Dafalias [110]

m_{S}	n_{S}	V
$0.05\alpha_{\mathrm{c}}^{\mathrm{c}}$	20	1000

Appendix D

Matlab Code for Constitutive models

D.1 Barodesy

D.1.1 Version of Medicus *et al.* for clay

```matlab
function [Tpv,ep] = getobjtr_baroclay(Tv,e,Dv,Wv,C)

%% Transform vector to matrix
T = v2m(Tv);
D = v2m(Dv);
W = v2ma(Wv);
getC;                          % extracts the material parameters

%% Calculation of normalised matrices
nrmD  = norm(D,'fro'); D0 = D/nrmD; trD0 = trace(D0);
nrmT  = norm(T,'fro'); T0 = T/nrmT; p    = -trace(T)/3;

%% critical void ratio
ec = exp(N-lambda*log(2*p))-1;

Lambda = -(lambda-kappa)/2/sqrt(3)*trD0+(lambda+kappa)/2;
beta   = -1/Lambda/c_3+2^(c_5*lambda)/sqrt(3)-1/sqrt(3);

R = getR(D,C); R0 = R/norm(R,'fro');

%% scalar quantities f, g and h
f = c_6*beta*trD0-1/2;
g = ((1+e)/(1+ec))^c_5+(1-c_6)*beta*trD0-1/2;
h = c_3 * nrmT;

%% Barodesy
Tp =  h*(f*R0+g*T0)*nrmD;

%% change of void ratio
edot =(1+e)*trace(D);
```

```
33  Tp = Tp+W*T—T*W;
34  Tpv = m2v(Tp);
35  ep = edot;
36  end
37
38  function R=getR(D,C)
39  getC;
40  zero = 10^—7;
41
42  %% Calculate constants
43  K_c   = (1—sind(phi_c))/(1+sind(phi_c));
44  K_0   = 1—sind(phi_c);
45  c_2   = (3*sqrt(K_c*(1—K_c)*K_0*(1—K_0))+3*K_c*(1—K_0))/(2*(K_c—K_0)));
46  c_1   = —K_c/(c_2^2*(K_c—1));
47
48  %% Calculate invariants of stretching
49  D0    = D/norm(D,'fro');
50  I_1   = D0(1,1)+D0(2,2)+D0(3,3);            % 1st invariant
51  I_2=1/2*(trace(D0)^2—trace(D0^2));          % 2nd invariant
52  J_2   = I_1^2/3—I_2+zero;                   % 2nd inv. of deviatoric  stretching
         tensor
53  D_dev = (3*J_2)^(1/2);                      % distortional stretching
54  m     = —I_1/(2/3*D_dev);                   % strain increment ratio
55
56  % R — function:
57  K     = —1/(1 + c_1*(m — c_2)^2)+1;
58  alpha = log(K+zero)/(D_dev+zero);
59  R     = —expm(alpha*D0);
60  end
```

D.1.2   Version of Kolymbas for sand

```
1   function [Tpv,Qpv] =getobjtr_baro2015(Tv,e,Dv,Wv,C)
2
3   %% Transform vectors in matrices
4   T = v2m(Tv);
5   D = v2m(Dv);
6   W = v2ma(Wv);
7
8   %% asign material constants
9   c_3=C(3);
10  c_4=C(4);
11  c_5=C(5);
12  e_c0=C(6);
13  e_min=C(7);
14
15  %% Calculate normalised matrices
16  nrmD  = norm(D,'fro'); D0 = D/nrmD; trD0 = trace(D0);
17  nrmT  = norm(T,'fro'); T0 = T/nrmT;
```

```
18  R = get_R(D,C); nrmR  = norm(R,'fro'); R0 = R/nrmR;
19
20  %% critical void ratio
21  B=(e_c0-e_min)/(e_c0+1)*((c_4+c_5*nrmT)/c_4)^(-(1+e_min)/c_5);
22  e_c=(e_min+B)/(1-B);
23  %% calculate f,g and h
24  f=trD0+c_3*e_c;
25  g=-c_3*e;
26  h=-(c_4+c_5*nrmT)/(e-e_min);
27
28  %% Barodesy
29  Tp= h*(f*R0+g*T0)*nrmD;
30
31  %% Change of void ratio
32  edot=(1+e)*trace(D);
33
34  Tp = Tp+W*T-T*W;
35  Tpv = m2v(Tp);
36  Qpv = edot;
37  end
38
39  function R=get_R(D,C)
40  %% Calculate the proportional stress path for the given D
41
42  %% asigning the variables
43  phi_c=C(1);
44
45  %% Calculating the constants
46  K_c=(1-sind(phi_c))/(1+sind(phi_c));      % stress_ratio in the critical
        state
47  c_1=sqrt(2/3)*log(K_c);
48  c_2=C(2);
49
50  %% Calculate the normalised D and the dilatancy
51  D0=D/norm(D,'fro'); trD0=trace(D0);
52
53  R=-expm(c_1*exp(c_2*trD0)*D0);
54  end
```

D.1.3   Improved version for sand

```
1  function [Tpv,Qpv] =getobjtr_v2018a1(Tv,Qv,Dv,Wv,C)
2
3  % Transformieren der Vektoren in Tensoren
4  T = v2m(Tv);
5  D = v2m(Dv);
6  e = Qv(1);
7  W = v2ma(Wv);
8  getC;                          % extracts the material parameters
```

```
 9
10   %% Calcultion of teh constants
11   A=(-T_r^alpha)/((E_r)*(1-alpha));
12   K_c=(1-sind(phi_c))/(1+sind(phi_c));
13   zeta=-1/K_c;
14   beta_komp=-(3+c_5*(1-c_e^zeta))/(sqrt(3));
15   beta_ext=(-3*kappa+c_5*(1-c_e^zeta))/sqrt(3);
16   c_8=-c_6/(beta_ext*sqrt(3));
17
18   %% Calculation of the normalised matrices
19   nrmD  = norm(D,'fro'); D0 = D/nrmD; trD0 = trace(D0);
20   nrmT  = norm(T,'fro'); T0 = T/nrmT;
21   R = get_R(D,C); nrmR  = norm(R,'fro'); R0 = R/nrmR;
22
23   %% Critical void ratio
24   e_c=(1+e_c0)*exp(A*trace(-T/3)^(1-alpha))-1;
25
26
27   %% scalar quantities f,g und h
28   h=E_r*(nrmT/sqrt(3)/T_r)^alpha;
29
30   beta=(beta_ext-beta_komp)/(2*sqrt(3))^c_7*(trD0+sqrt(3))^c_7+beta_komp;
31   f=c_8*beta*trD0+c_6;
32   g=(1-c_8)*beta*trD0+c_5*(((1+e)/(1+e_c))^zeta-1)-c_6;
33
34   %% Barodesy
35   Tp=h*(f*R0+g*T0)*nrmD;
36
37   %% change of void ratio
38   edot=(1+e)*trace(D);
39
40   Tp = Tp+W*T-T*W;
41   Tpv = m2v(Tp);
42   Qpv = edot;
43   end
44
45   function R = get_R(D,C)
46
47   phi_c=C(1);
48   %% Calculate the stress ratio at given states
49   K_c   = (1-sind(phi_c))/(1+sind(phi_c));
50   K_0   = 1-sind(phi_c);
51   K_p   = (9-11*sind(phi_c))/(9+13*sind(phi_c));
52
53   x=sqrt(2);                              % dilatancy at maximum eta
54   alpha_pres=ones(4,1)*NaN;
55   delta_pres=alpha_pres;
56
57   delta_pres(1)=-1;                       % oedometric compression
58   alpha_pres(1)=log(K_0);
59   delta_pres(2)=0;                        % critical state
```

```
60  alpha_pres(2)=sqrt(2/3)*log(K_c);
61  delta_pres(3)=sqrt(3)/3;                    % peak state
62  alpha_pres(3)=log(K_p)/sqrt(3/2-1/6);
63
64  %% Determine the constants
65  c_min=-30;                                  % alpha_min
66  c_1=(log((1+x)/x)*log((alpha_pres(3)-c_min)/(alpha_pres(2)-c_min))-...
67      log((alpha_pres(1)-c_min)/(alpha_pres(2)-c_min))*log((3*x-sqrt(3))/(3*x)))/
        ...
68      (log((x+1)/x)*log((3+3*x)/(3+3*x-sqrt(3)))-log((1+x)/(2+x))*...
69      log((3*x-sqrt(3))/(3*x)));
70
71  c_2=(log((alpha_pres(1)-c_min)/(alpha_pres(2)-c_min))-c_1*log((1+x)/...
72      (2+x)))/log((1+x)/x);
73  c_3=(alpha_pres(2)-c_min)*(1+x)^c_1/x^c_2;
74
75  %% Calculate the normalised D and the dilatancy
76  D0 = D/norm(D,'fro');
77  delta=trace(D0);
78
79  alpha=c_min+c_3*abs(delta-sqrt(2))^c_2/(1+abs(delta-sqrt(2)))^c_1;
80  R = -expm(alpha*D0);
81  end
```

D.2   Hypoplasticity

D.2.1   Version of von Wolffersdorff for sand

```
1   function [Tdot,edot] = getobjtr_hypo(T,e,D,W,C)
2
3   % Hypoplasticity: Version Gudehus/Wolffersdorff/Herle 1995
4   % W. Fellin, Okt. 2007
5   %
6   % input: (T,D,W,e,C)
7   % state variables
8   % T ... stress tensor as vector (6,1)
9   % e ... void ratio
10  % deformation
11  % D ... streching tensor as vector (6,1)
12  % W     spin tensor as vector (6,1)
13  % material constants
14  %          C(1)      phi_c
15  %          C(2)      h_s
16  %          C(3)      n
17  %          C(4)      e_d0
18  %          C(5)      e_c0
19  %          C(6)      e_i0
20  %          C(7)      alpha
```

```
21   %          C(8)        beta
22   % output: [Tdot,edot]
23   % Tdot ... stress rate tensor as vector (6,1)
24   % edot ... time rate of void ratio
25
26   % T=[Tvec(1) Tvec(4) Tvec(5)
27   %     Tvec(4) Tvec(2) Tvec(6)
28   %     Tvec(5) Tvec(6) Tvec(3)];
29   %
30   % D=[Dvec(1) Dvec(4) Dvec(5)
31   %     Dvec(4) Dvec(2) Dvec(6)
32   %     Dvec(5) Dvec(6) Dvec(3)];
33   %
34   % W=[Wvec(1) Wvec(4) Wvec(5)
35   %    -Wvec(4) Wvec(2) Wvec(6)
36   %    -Wvec(5) -Wvec(6) Wvec(3)];
37
38   I = eye(3,3);
39
40   phic = C(1)*pi/180;
41   hs   = C(2);
42   n    = C(3);
43   ed0  = C(4);
44   ec0  = C(5);
45   ei0  = C(6);
46   alpha = C(7);
47   beta  = C(8);
48
49   trD = trace(D);
50   valD = sqrt(trace(D^2));
51   trT = trace(T);
52
53   % for zero (or positive) stress ---> hypoplastic compression law
54   tolabs = 1e-6;
55   if (abs(trT) < tolabs) || (trT > 0)
56       trT
57       To = zeros(3,3);
58       if ((trD<0) && (e < ei0))
59           % derivative of compression law
60           edot = ( 1 + e )*trD;
61           To(1,1) = edot/e/n/3 * tolabs / ( tolabs/hs )^n;
62           To(2,2) = To(1,1);
63           To(3,3) = To(1,1);
64           Tdot = To + W*T - T*W;
65           Tdotvec=[Tdot(1,1); Tdot(2,2); Tdot(3,3); Tdot(1,2); Tdot(1,3); Tdot
                        (2,3)];
66       end
67       return
68   end
69
70
```

```
71  sinphi = sin(phic);
72  sq2     = sqrt(2);
73  sq3     = sqrt(3);
74  sq6     = sqrt(6);
75
76  % relativ stress: Ts = T / trace(T)
77  if ( abs(trT) < 1.e-10 )
78      Ts = T./1e-10;
79  else
80      Ts = T ./ trT;
81  end
82
83  % square of relativ stress: Ts2 = Ts * Ts
84  Ts2 = Ts^2;
85  trTs2 = trace(Ts2);
86
87  % deviator: Tsv = Ts - 1/3 I
88  Tsv = Ts - 1/3*I;
89  Tsv2 = Tsv*Tsv;
90  Tsv3 = Tsv2*Tsv;
91  trTsv2 = trace(Tsv2);
92  trTsv3 = trace(Tsv3);
93
94  % norm of relativ deviator stress, ||Tsv||
95  valTsv = sqrt(trace(Tsv2));
96
97  if ( trTsv2 < 1.e-10 )
98    c3t = 1;
99  else
100   c3t = -sq6*trTsv3/trTsv2^1.5;
101   if ( c3t > 1 )
102     c3t =  1;
103   end
104   if ( c3t < -1 )
105     c3t = -1;
106   end
107 end
108
109
110 tpsi  = sq3*valTsv;
111 tpsi2 = tpsi^2;
112
113
114 term1 = tpsi2/8 + ( 2 - tpsi2 )/( 2 + sq2*tpsi*c3t );
115 if term1 < 0
116   Fm = 1e-10;
117 else
118   term1 = sqrt(term1);
119 end
120 term2 = tpsi/2/sq2;
121 Fm  = term1-term2;
```

```
122
123  Fm2 = Fm^2;
124
125  a    = sq3*( 3 - sinphi )/( 2*sq2*sinphi );
126  a2   = a^2;
127
128  ed = ed0*exp(-(-trT/hs)^n);
129  ec = ec0*exp(-(-trT/hs)^n);
130  ei = ei0*exp(-(-trT/hs)^n);
131
132  term3 = ( (ei0-ed0)/(ec0-ed0) )^alpha;
133  term4 = 1 / ( 3 + a2 - a*sq3*term3 );
134  term5 = hs/n*( ei0/ec0 )^beta * ( 1+ei )/ei*( -trT/hs )^(1-n);
135
136  fb  = term4*term5;
137  fe  = ( ec/e )^beta;
138
139  fd = ( e-ed )/( ec-ed );
140  if fd > 0
141     fd = fd^alpha;
142  else
143     fd = 0;
144  end
145
146  fss = fb*fe/trTs2;
147
148  % objective time rate
149  To = fss*(Fm2*D + a2*trace(Ts*D)*Ts + fd*a*Fm*(Ts+Tsv)*valD);
150
151  % time rate
152  if norm(W)>0
153     Tdot = To + W*T - T*W;
154  else
155  Tdot = To;
156  end
157
158  edot = ( 1 + e )*trD;
```

D.2.2   Intergranular strain for the version of von Wolffersdorff

```
1  function [Tdot,Qvecdot] = getobjtr_hypo_id(T,Qvec,D,W,C)
2  % Hypoplasticity: Version Niemunis/Herle 1997
3  % W. Fellin, M. Mittendorfer, Okt. 2007
4  %
5  % input: (T,D,W,Q,C)
6  % state variables
7  % Tvec ... stress tensor as vector (6,1)
8  % Qvec(1) ... void ratio
9  % Qvec(2:7) ... intergranular strain as vector
```

```
10  % deformation
11  % Dvec ... streching tensor as vector (6,1)
12  % Wvec ... spin tensor as vector (6,1)
13  % material constants
14  %          C(1)       phi_c
15  %          C(2)       h_s
16  %          C(3)       n
17  %          C(4)       e_d0
18  %          C(5)       e_c0
19  %          C(6)       e_i0
20  %          C(7)       alpha
21  %          C(8)       beta
22  %          C(9)       R
23  %          C(10)      mR
24  %          C(11)      mT
25  %          C(12)      beta_r
26  %          C(13)      chi
27  % output: [Tdot,Qdot]
28  % Tvecdot ... stress rate tensor as vector (6,1)
29  % Qvecdot(1,1) ... time rate of void ratio
30  % Qvecdov(2:7,1) ... time rate of ntergranular strain
31
32  I = eye(3,3);
33
34  phic = C(1)*pi/180;
35  hs   = C(2);
36  n    = C(3);
37  ed0  = C(4);
38  ec0  = C(5);
39  ei0  = C(6);
40  alpha = C(7);
41  beta  = C(8);
42  R = C(9);
43  mR = C(10);
44  mT = C(11);
45  beta_r = C(12);
46  chi = C(13);
47
48  T = v2m(Tvec);
49  D = v2m(Dvec);
50  W = v2ma(Wvec);
51  e = Qvec(1);
52  d = v2m(Qvec(2:7));
53
54  D_D = tensprod2dd2(D,D);
55  valD = sqrt(D_D);
56
57  trD = trace(D);
58  trT = trace(T);
59
60  if trT >= 0
```

```
 61        Qvecdot(1) = (1 + e)*trD;
 62        Tdot = zeros(3,3);
 63        Tvecdot = m2v(Tdot);
 64   else
 65        % relativ stress: Th = T / trace(T)
 66        Th = T ./ trT;
 67
 68        % deviator: Ths = Th - 1/3 I
 69        Ths = Th - 1/3*I;
 70
 71        % norm of relativ deviator stress, ||Ths||
 72        Ths_Ths = tensprod2dd2(Ths,Ths);
 73        valThs = sqrt(Ths_Ths);
 74
 75        tpsi  = sqrt(3)*valThs;
 76
 77        ThsThs = dotprod22(Ths,Ths);
 78        ThsThsThs = dotprod22(ThsThs,Ths);
 79
 80        if Ths_Ths < 1.e-10
 81            c3t = 1;
 82        else
 83            c3t = -sqrt(6)*trace(ThsThsThs)/(Ths_Ths)^1.5;
 84            if c3t > 1
 85                c3t = 1;
 86            elseif c3t < -1
 87                c3t = -1;
 88            end
 89        end
 90
 91        termF = tpsi^2/8 + (2-tpsi^2)/(2+sqrt(2)*tpsi*c3t);
 92        if termF < 0
 93            F = 1e-10;
 94        else
 95            F = sqrt(termF)-tpsi/(2*sqrt(2));
 96        end
 97
 98        a = sqrt(3)*(3-sin(phic))/(2*sqrt(2)*sin(phic));
 99
100        ed = ed0*exp(-(-trT/hs)^n);
101        ec = ec0*exp(-(-trT/hs)^n);
102        ei = ei0*exp(-(-trT/hs)^n);
103
104        term1 = ((ei0-ed0)/(ec0-ed0))^alpha;
105        term2 = 1/(3 + a^2 - a*sqrt(3)*term1);
106        term3 = hs/n * (ei/e)^beta * ((1+ei)/ei) * (-trT/hs)^(1-n);
107
108        fbfe = term2*term3;
109        fd = ((e-ed)/(ec-ed))^alpha;
110
111        II = zeros(3,3,3,3);
```

```
112        for i=1:3
113            for j=1:3
114                II(i,j,i,j) = 1;
115            end
116        end
117
118     ThTh = tensprod22(Th,Th);
119     ThddTh = tensprod2dd2(Th,Th);
120      for ii=1:3
121            for jj=1:3
122                for kk=1:3
123                    for ll=1:3
124                        LL(ii,jj,kk,ll) = fbfe/ThddTh*(F^2*II(ii,jj,kk,ll)...
125                            + a^2*ThTh(ii,jj,kk,ll));
126                    end
127                end
128                NN(ii,jj) = fd*fbfe*F*a/ThddTh*(Th(ii,jj) + Ths(ii,jj));
129            end
130        end
131
132     % calculate  TR_ij = L_ijkl*D_kl + N_ij*||D||
133     TR1 = tensprod4dd2(LL,D) + NN*valD;
134
135     nd = norm(d,'fro');
136     if R < 1e-12
137         rho = 1;
138     else
139         rho = nd/R;
140     end
141
142     if nd > 0
143         dh = d/nd;
144     else
145         dh = zeros(3:3);
146     end
147
148     LL_dh = tensprod4dd2(LL,dh);
149     LLdhdh = tensprod22(LL_dh,dh);
150     NNdh = tensprod22(NN,dh);
151     dh_D = tensprod2dd2(dh,D);
152
153     for ii=1:3
154         for jj=1:3
155             for kk=1:3
156                 for ll=1:3
157                     if dh_D > 0
158                         MM(ii,jj,kk,ll) = (rho^chi*mT + (1-rho^chi)*mR)*LL(ii,
159                             jj,kk,ll)...
                             + rho^chi*(1-mT)*LLdhdh(ii,jj,kk,ll) + rho^chi*NNdh
                                 (ii,jj,kk,ll);
160                     else
```

```
161                          MM(ii,jj,kk,ll) = (rho^chi*mT + (1-rho^chi)*mR)*LL(ii,
                                 jj,kk,ll)...
162                              + rho^chi*(mR-mT)*LLdhdh(ii,jj,kk,ll);
163                     end
164                  end
165               end
166           end
167       end
168
169       % calculate TR_ij = M_ijkl : D_kl
170       TR2 = tensprod4dd2(MM,D);
171
172       TR = TR2;
173
174       % objective time rate of intergarnular strain
175       if tensprod2dd2(dh,D) > 0
176           dR = tensprod4dd2(II-tensprod22(dh,dh)*rho^beta_r,D);
177       else
178           dR = D;
179       end
180
181       % time rate of stress and intergarnular strain
182           Tdot = TR +W*T-W*T;
183           ddot = dR;
184
185       Tvecdot = m2v(Tdot);
186       Qvecdot(2:7,1)=m2v(ddot);
187
188       % time rate of void ratio
189       Qvecdot(1) = (1 + e)*trD;
190
191 end
192
193 end % hypo_id
194
195
196 % doble dot product of the second order tensors T and D
197 function TD = tensprod2dd2(T,D)
198 % T:D = T_ij D_ij
199 TD = sum(sum(T.*D));
200 end
201
202 % dot product of the second order tensors T and D
203 function TD = dotprod22(T,D)
204 % T.D = T_ij D_jk
205 TD = T*D;
206 end
207
208 % doble dot product of the fourth order tensor T with the second order tensor D
209 function LD = tensprod4dd2(L,D)
210 % L:D = L_ijkl Dkl
```

```
211
212   for i = 1:3
213       for j = 1:3
214           Lij=reshape(L(i,j,:,:),3,3);
215           LD(i,j) = sum(sum(Lij.*D));
216       end
217   end
218
219   end
220
221   % tensor product of two second order tensors
222   function TD = tensprod22(T,D)
223   % TD_ijkl = Tij D_kl
224
225   for i=1:3
226     for j=1:3
227       for k=1:3
228         for l=1:3
229           TD(i,j,k,l)=T(i,j)*D(k,l);
230         end
231       end
232     end
233   end
234
235   end
```

D.3   Sanisand

```
1    function [D,dE] = strain(T,E,parms,dTo)
2    %% STRAIN Calculate the strain from stress change
3
4    %% assigning of variables
5    dsigma=v2m(dTo);                    % convert vector to matrix
6    sigma=v2m(T);                       % convert vector to matrix
7
8    p0=E(1);
9    alpha=[E(2) E(5) E(6); E(5) E(3) E(7); E(6) E(7) E(4)];
10   void=E(8);
11
12   getparms                            % extract paramters from parms
13
14   %% calculate mean and deviatoric stress
15   dp= trace(dsigma)/3;
16   ds = dsigma-eye(3,3)*dp;
17   p= trace(sigma)/3;
18   s = sigma-eye(3,3)*p;
19
20   %% elastic strain
```

```
21  ratio = (3/2)*(1—2*nu)/(1+nu);
22  K0    = G0/ratio;
23  fe  = (2.97—void)*(2.97—void)/(1+void);
24  fe1 = (1+void)/void;
25
26  G = G0*p_a*fe*sqrt(p/p_a);
27  K  = K0*p_a*fe1*((p/p_a)^(2/3));
28
29  epsve=dp/K;
30  epsee=ds/(2*G);
31
32  %% yield function
33  sig_tr=sigma+dsigma;                    % trial stress state
34  p_tr=trace(sig_tr)/3;
35  s_tr=sig_tr—p_tr*eye(3,3);
36
37  % Calculate the distance between trial state and back stress axis
38  x=s_tr—p_tr*alpha;
39  contr=0;
40  for i=1:3
41      for j=1:3
42          contr=contr+x(i,j)*x(i,j);
43      end
44  end
45  f = contr*3/2—(0.05*alpha_cc)^2*p_tr^2*(1—(p_tr/p0)^20);
46
47  %% check if yielding
48  if f<—1e—12 % elastic range
49      dEps=epsee+epsve/3*eye(3,3);
50      D=m2v(dEps);
51      dE=zeros(10,1);
52      dE(8)=—(1 + void)*epsve;          % change in void ratio
53      dE(9)=0;                          % No yielding
54      dE(10)=f;                         % return the actual value of the yield
               funct.
55      return
56  elseif E(9)==0 && f<0 % unloading
57      dEps=epsee+epsve/3*eye(3,3);
58      D=m2v(dEps);
59      dE=zeros(10,1);
60      dE(8)=—(1 + void)*epsve;
61      dE(9)=0;
62      dE(10)=f;
63      return
64  elseif E(9)==0 % partial elastic, partial plastic
65      %% besectional algorithm to find f(dsig)=0
66      x0 = 0;
67      x1 = 1;
68      xm=.5;
69      k=1;
70      while 1
```

```
71          sig_tr=sigma+dsigma*xm;
72          p_tr=trace(sig_tr)/3;
73          s_tr=sig_tr-p_tr*eye(3,3);
74          x=s_tr-p_tr*alpha;
75          contr=0;
76          for i=1:3
77              for j=1:3
78                  contr=contr+x(i,j)*x(i,j);
79              end
80          end
81          f = contr*3/2-(0.05*alpha_cc)^2*p_tr^2*(1-(p_tr/p0)^20);
82          if abs(f)<1e-11
83              break
84          elseif f<0
85              x0=xm;
86          else
87              x1=xm;
88          end
89          xm = x0 + (x1-x0)/2;
90      end
91      dp=(1-xm)*dp;
92      ds=(1-xm)*ds;
93      T=sigma+dsigma*xm;
94      p=trace(T)/3;
95      s=T-p*eye(3,3);
96  end
97
98  %% plastic strain
99
100 % Lode angle
101 r=s/p;
102 nenner=0;
103 for i=1:3
104     for j=1:3
105         nenner=nenner+(r(i,j)-alpha(i,j))*(r(i,j)-alpha(i,j));
106     end
107 end
108 n=(r-alpha)/sqrt(nenner);
109 trn3=trace(n^3);
110 theta3=acos(sqrt(6)*trn3)/3;
111 if not(isreal(theta3))
112     theta3=acos(trn3/abs(trn3))/3;
113 end
114 % Value of the critical surface
115 g=2*c_alpha/((1+c_alpha)-(1-c_alpha)*cos(3*theta3));
116
117 % critical void ratio
118 ec = e0 - lambda*(p/p_a)^xi;
119 psi = void - ec;
120
121 % opening angle alpha^b und alpha^d
```

```matlab
alpha_cb=alpha_cc*exp(-n_b*psi);
alpha_cd=alpha_cc*exp(n_d*psi);

% alpha_bar
b0=G0*h0*(1-c_h*void)*sqrt(p_a/p);
b_ref=alpha_cb*(1+c_alpha);

nenner=0;
for i=1:3
    for j=1:3
        nenner=nenner+(sqrt(2/3)*b_ref*n(i,j)-(sqrt(2/3)*g*alpha_cb*n(i,j)...
            -alpha(i,j)))*n(i,j);
    end
end
h=b0/(3/2*nenner^2);
r_ef=(0.05*alpha_cc)*sqrt(1-(p/p0)^20);

alpha_bar=h*r_ef*(sqrt(2/3)*g*alpha_cb*n-alpha);

%p0_bar
p_b=p_r*void^(-1/rho_c);
alpha_s=0;
for i=1:3
    for j=1:3
        alpha_s=alpha_s+alpha(i,j).^2;
    end
end
if alpha_s == 0
    nalpha=zeros(3,3);
else
    nalpha=alpha/sqrt(alpha_s);
end
trna3=trace(nalpha^3);
thetaalpha=acos(sqrt(6)*trna3)/3;
if not(isreal(thetaalpha))
    thetaalpha=acos(trna3/abs(trna3))/3;
end

galpha=2*c_alpha/((1+c_alpha)-(1-c_alpha)*cos(3*thetaalpha));

delta=1-p/p_b*(1+3*alpha_s/(galpha*alpha_cc).^2);
if abs(delta)<1e-6
    sgn=0;
else
    sgn=delta/abs(delta);
end

p0_bar=p0*(1+void)*exp(-1000*r_ef)/(void*(rho_c-(p0/p_a)^(1/3)/K0)*...
    (1-sgn*abs(delta)^theta));

%plastic modulus
```

```
173    dfdalpha=-3*p*(s-p*alpha);
174    dfdp0=-20*(0.05*alpha_cc)^2*p^2/p0*(p/p0)^20;
175
176    Kp=0;
177    for i=1:3
178        for j=1:3
179            Kp=Kp+dfdalpha(i,j)*alpha_bar(i,j);
180        end
181    end
182    Kp=-(Kp+dfdp0*p0_bar);
183
184
185    dfds=3*(s-p*alpha);
186    dfdp=0;
187    for i=1:3
188        for j=1:3
189            dfdp=dfdp+alpha(i,j)*(s(i,j)-p*alpha(i,j));
190        end
191    end
192    dfdp=-3*dfdp-2*(0.05*alpha_cc)^2*p+(2+20)*(0.05*alpha_cc)^2*p*(p/p0)^20;
193
194    % plasitc multiplier
195    L=0;
196    for i=1:3
197        for j=1:3
198            L=L+dfds(i,j)*ds(i,j);
199        end
200    end
201    L=L+dfdp*dp;
202    L=L/Kp;
203
204
205    dalpha=L*alpha_bar;
206    dp0=L*p0_bar;
207
208    % plastic strain
209    x=0;
210    for i=1:3
211        for j=1:3
212            x=x+(sqrt(2/3)*g*alpha_cd*n(i,j)-alpha(i,j))*n(i,j);
213        end
214    end
215    D=sqrt(3/2)*A_d*x;
216
217    epsvp=L*(D*r_ef+exp(-1000*r_ef));
218    epsep=L*(sqrt(3/2)*n*r_ef+3/2*Chi*r*exp(-1000*r_ef));
219
220    % calculate the new value of the yield function
221    sig_tr=sigma+dsigma;
222    p_tr=trace(sig_tr)/3;
223    s_tr=sig_tr-p_tr*eye(3,3);
```

```
224  x=s_tr—p_tr*(alpha+dalpha);
225  contr=0;
226  for i=1:3
227      for j=1:3
228          contr=contr+x(i,j)*x(i,j);
229      end
230  end
231  f = contr*3/2—(0.05*alpha_cc)^2*p_tr^2*(1—(p_tr/(p0+dp0)))^20);
232
233  epsv=epsve+epsvp;
234  epse=epsee+epsep;
235
236  dEps=epse+epsv/3*eye(3,3);
237  D=m2v(dEps);
238  dE=zeros(10,1);
239  dE(1)=dp0;
240  dE(2:7)=m2v(dalpha);
241  dE(8)=—(1 + void)*epsv;
242  dE(9)=1;
243  dE(10)=f;
244  end
```

Appendix E

Matlab Code for Stress rotation

E.1 Stress dependant

```matlab
1  %% Calculating  stress rotation for Sanisand
2
3  setparms;                              % set material parameters,
4  setinit;               % set initial condition for stress T and internal
       variables E,
5  omega=-10*pi;                          % number of rotations
6
7  T_0=T;                                 % Principal stress tensor
8  Dt=.5e-5;
9  t=0;
10 i=1;
11 x=zeros(4,1);                          % value of the deformation function
12 xp=zeros(4,1);                         % time derivative of x
13
14 tdes=1;
15
16 while t<tdes
17     dT_des=[T_0(1)+(T_0(1)-T_0(2))/2*(-1+cos(2*omega*(t+Dt)));...
18         T_0(2)-(T_0(1)-T_0(2))/2*(-1+cos(2*omega*(t+Dt)));T_0(3);...
19         -(T_0(1)-T_0(2))/2*sin(2*omega*(t+Dt));0;0]-T; % stress increment
20     n=1;
21
22     dTo=dT_des;                        % Estimation for the first step
23     dT=zeros(6,1);
24     R=dT_des-dT;
25     while norm(R)>1e-6                 % Iterate as long as the residuum is to
           large
26         [dT,k]=newton(T,E,parms,dTo,x);   % Get the stress rateand
27                                        % and the Jacobian
28         R=dT-dT_des;
29
30         dTo(1:4)=dTo(1:4)-k(1:4,1:4)\R(1:4);% New estimation of dTo
31         n=n+1;
32     end
33     [D,dE]=strain(T,E,parms,dTo);          % calculate D and the rate of E
34
35     xp(3)=(1+x(3))*D(1);                   %calculate the time derivative of x
```

```matlab
36        xp(2)=(1+x(2))*D(2);
37        xp(4)=(1+x(4))*D(3);
38        xp(1)=1/(1+x(3))*(2*D(4)*(1+x(3))*(1+x(2))+xp(3)*x(1));
39
40        E(9:10)=0;
41
42        %% Update the variables
43        E=E+dE;
44        T=T+dT;
45        i=i+1;
46        x=x+xp;
47
48        t=t+Dt;
49   end
```

```matlab
1    function [dT,k]=newton(T,E,parms,dTo,x)
2    %% Calculate the Jacobian and the stress rate for Sanisand
3
4    k=ones(6,4)*NaN;                    % Initiate Jacobian
5
6    theta=sqrt(eps);                    % variation
7
8    [D,~]=strain(T,E,parms,dTo);        % calculate stretching for the given stress
            rate
9
10   %% Determine time derivative of x from stretching
11   xp(3)=(1+x(3))*D(1);
12   xp(2)=(1+x(2))*D(2);
13   xp(4)=(1+x(4))*D(3);
14   xp(1)=1/(1+x(3))*(2*D(4)*(1+x(3))*(1+x(2))+xp(3)*x(1));
15
16   W=1/(2*(1+x(2))*(1+x(3)))*[0;0;0;xp(1)*(1+x(3))-xp(3)*x(1);0;0];
17
18   %% Transform vectors [6x1] (Voigt-Notation) to matrices [3x3]
19   Wmat=v2ma(W);
20   Tmat=v2m(T);
21   dTomat=v2m(dTo);
22
23   %% Calculate stress rate
24   dTmat=dTomat+Wmat*Tmat-Tmat*Wmat;
25
26   dT=m2v(dTmat);
27
28   %% Calculate the stress derivatives on dTo
29   for i=1:4
30       V=zeros(6,1);
31       V(i)=1;
32       [Dq,~]=strain(T,E,parms,dTo+theta*V);   % variation of the i-th component
33
34       % Determine time derivative of x from stretching
35       xp(3)=(1+x(3))*Dq(1);
```

```
36      xp(2)=(1+x(2))*Dq(2);
37      xp(4)=(1+x(4))*Dq(3);
38      xp(1)=1/(1+x(3))*(2*Dq(4)*(1+x(3))*(1+x(2))+xp(3)*x(1));
39
40      W=1/(2*(1+x(2))*(1+x(3)))*[0;0;0;xp(1)*(1+x(3))-xp(3)*x(1);0;0];
41
42      Wmat=v2ma(W);
43      Tmat=v2m(T);
44      dTomat=v2m(dTo+theta*V);
45
46      % Calculate stress rate
47      dTmat=dTomat+Wmat*Tmat-Tmat*Wmat;
48
49      dTq=m2v(dTmat);
50
51      k(:,i)=1/theta*(dTq-dT);        % i-th column of the Jacobian
52  end
53  end
54
55  %% convert a vector to a symmetric matrix
56  function M=v2m(V)
57  M = [V(1,1),V(4,1),V(5,1);V(4,1),V(2,1),V(6,1);V(5,1),V(6,1),V(3,1)];
58  end
59
60  %% convert a matrix to a vector
61  function v=m2v(M)
62  v = [M(1,1),M(2,2),M(3,3),M(1,2),M(1,3),M(2,3)]';
63  end
64
65  %% convert a vector to an asymmetric matrix
66  function M=v2ma(V)
67  M = [V(1,1),V(4,1),V(5,1);-V(4,1),V(2,1),V(6,1);-V(5,1),-V(6,1),V(3,1)];
68  end
```

E.2   Strain dependant

```
1   %% Calculating stress rotation for Hypoplasticity and Barodesy
2
3   setC;                              % set Material parameters
4   setInit;                           % set initial conditions for stress and internal
        variables
5   omega=-10*pi;
6
7   T_0=T;                             % Principal stress tensor
8   Dt0=1e-5;
9   Dt=Dt0;
10
11  t=0;
```

```matlab
12  tdes=1.5;
13  i=1;
14  x=zeros(4,1);                      % current value of the deformation
        functions
15  xp=[1;-1;.5;0];                    % first estimation of the time derivative
        of x
16  dof=[1,2,3,4];                     % Degrees of freedom for stress
17  dofx=1:4;                          % Degrees of freedom in deformation
18  while t<tdes
19      Dt=Dt0;
20      T_des=[T_0(1)+(T_0(1)-T_0(2))/2*(-1+cos(2*omega*(t+Dt)));...
21          T_0(2)-(T_0(1)-T_0(2))/2*(-1+cos(2*omega*(t+Dt)));T_0(3);...
22          -(T_0(1)-T_0(2))/2*sin(2*omega*(t+Dt));0;0];
23      R=ones(6,1);                   % Initiating the Residual
24
25      while (norm(R(dof),2)>1e-12)   % Newton iteration to get the correct xp
26          [Ts,~,k,~]=Umat_nl(T,E,x,xp,Dt);
27
28          K=reshape(k,6,4)*Dt;       % Jacobian
29          R=Ts-T_des;                % Residual for the current xp
30          xp(dofx)=xp(dofx)-K(dof,dofx)\R(dof);    % Estimation for the new xp
31      end
32
33      [T,E,~,x]=Umat_nl(T,E,x,xp,Dt);
34
35      i=i+1;
36      t=t+Dt;
37  end
```

```matlab
1   function [T,E,varargout]=Umat_nl(T,E,x,xp,Dt,C)
2   %% Umat_nl returns the stress T, internal variables E and the Jacobian for
3   %% after a given timestep and deformation derivative
4   %% T stress
5   %% E internal variables
6   %% vargout contains the Jacobian and the new value of the deformation functions
7   %% x vector with the value of the deformation function
8   %% xp derivative of x
9   %% Dt timestep
10  %% C Material parameters
11
12  global material_model
13
14  nout=max(nargout,1)-2;
15
16  theta=sqrt(eps);                   % Variation
17
18  m=length(E);
19  y=zeros(6*5+5*m,1);                % Initiation of the solution vector
20  y(1:6)=T;                          % assign current stress state
21  y(5*6+1:5*6+m)=E;                  % assign current internal variables
22  doty=ones(size(y))*NaN;
```

```
23   dotx=ones(size(y))*NaN;
24
25   tr=str2func(['getobjtr_',material_model]);          % Choose the material model
26
27   D=[xp(3)/(1+x(3));xp(2)/(1+x(2));xp(4)/(1+x(4));...        %Calculate D and W
28       (xp(1)*(1+x(3))-xp(3)*x(1))/(2*(1+x(2))*(1+x(3)));0;0];
29   W=1/(2*(1+x(2))*(1+x(3)))*[0;0;0;xp(1)*(1+x(3))-xp(3)*x(1);0;0];
30
31   [doty(1:6),doty(31:30+m)] = tr(T,E,D,W,C); % Calculate the current stress rate
         and rate of the internal variables
32
33   if nout==2
34       for i=1:4                % Variation of each component in xp for Jacobian
35           V=zeros(4,1);
36           V(i)=1;
37           xpq=xp+theta*V;
38           D=[xpq(3)/(1+x(3));xpq(2)/(1+x(2));xpq(4)/(1+x(4));...
39               (xpq(1)*(1+x(3))-xpq(3)*x(1))/(2*(1+x(2))*(1+x(3)));0;0];
40           W=1/(2*(1+x(2))*(1+x(3)))*[0;0;0;xpq(1)*(1+x(3))-xpq(3)*x(1);0;0];
41
42           [Tdq,Qdq]=tr(T,E,D,W,C); %
43
44           doty(6*i+1:6*(i+1))=1/theta*(Tdq-doty(1:6));
45           doty(31+m*i:30+m*(i+1))=1/theta*(Qdq-doty(31:30+m));
46       end
47   end
48
49   u=y+Dt*doty;
50   T=u(1:6);
51   E=u(31:30+m);
52   if nout>0
53       varargout{nout}=x+xp*dt;
54   end
55   if nout==2
56       varargout{1}=1/Dt*y(7:30);          % Calculates the Jacobian
57   end
58   end
```